BRAIN-CENTRIC

BRAIN-CENTRIC

BRAIN-CENTRIC

How the mental space builds realities

Part one: THE MENTAL SPACE

BRAIN-CENTRIC

BRAIN-CENTRIC

How the mental space builds realities

Part one: THE MENTAL SPACE

RONALD CICUREL

By the same author

Spyridon's Quest, 2001. Editions Sarina

L'ordinateur ne digérera pas le cerveau, science et cerveaux artificiels, essai sur la nature du Réel 2013, Editions Sarina

The Relativistic Brain: How it Works and why it Cannot be Simulated by a Turing Machine
in collaboration with Miguel Nicolelis (English and Portuguese versions) 2015 Kios Publishing

Nausées, danses avec l'irréel,
(French) 2016, Editions Sarina

Mémoires du Caire: Souvenirs d'enfance d'un grand-père juif d'Egypte
(French Edition), 2018, Editions Sarina

La réalité est une création humaine.
Reality is a human construct.
(French and English versions) 2021 Editions Sarina

Que devient la vérité si tout se passe dans notre cerveau, De l'analogue au digital. (French) Editions Sarina 2021

ooooooo

ooooooo

Brain-centric, How the Mental Space Builds Realities.
Part one: **The Mental Space. With 39 drawings.**
© 2020 Ronald Cicurel, éditions Sarina.

editions-sarina.ch

Second Edition, February 2022

ISBN: 9781082275562

Part one: THE MENTAL SPACE

Thank you

To my children Valerie, Samuel, and Sarina and my grandson Ben, that made me a proud father and grandfather. With all my love, respect, and esteem for my so devoted companion, Liliane. My profound gratitude for supporting and encouraging me through good and bad weather.

With all my gratitude to my friend Professor Miguel Nicolelis for the years-long passionate discussions, days, and nights in Lausanne, Montreux, Natal, Sao Paulo, and other places on this planet. Thank you with all my heart, Miguel, for sharing your ideas and inspiring mine.

And to my friends and colleagues for sharing with me and letting me burrow and adopt their ideas. Admiring other humans is a privilege. A special thank you to Lecia Caris that I respect for the logic and creativity of her writing. She motivated me and helped me with her remarks complete the redaction of this book. I am also grateful to Hélène Potier for her comments and proofreading.

I am grateful to all the brilliant thinkers and scientists that allow us to make some sense of our human existence and set an example. It cited most of them in the text.

I feel much closer to scientists up to the 1960s. Is it because of my age or some other mental biases? But I feel these people were thinkers or researchers more than scientists. Most were not part of any organized system and free to go in the direction they had chosen. They were not running as much as now after celebrity, publications, hype, and honors. Their passion was the quest for truth, and they did not hesitate to take risks and step out of their primary competences. Heroes are not the people who have the power; heroes are the people who contest power. Heroes are not professionals, they are rather passionate amateurs.

CONTENTS

Preface 13
History, why this book 15
Notes to the reader 18

I: PERCEPTION

1.1 Survival 21
1.2 Perception 23
1.3 Expectations 31
1.4 Filling in 34
1.5 Individuation 37
1.6 Fragmentation 40
1.7 Mental worlds 42
1.8 Two information sources 47
1.9 Are expectations universal? 49

II: THE FABRIC OF THE MENTAL SPACE

2.1 Memory 53
2.2 Brains and mental spaces 58
2.3 Computer's mental space 61
2.4 Mental Patterns 64
2.5 Mental abstractions 68
2.6 Orders and invisible worlds 72
2.7 Cave painting 74
2.8 Cybernetics, adaptation 78
2.9 Adaptation, Complexity and Russian dolls 85
2.10 Extraterrestrial intelligence 90
2.11 Life and Survival are part of Intelligence 95
2.12 Curiosity and Learning 98

III: THE VERBAL WORLD

3.1 Languages and mental spaces 103
3.2 Maps and territories 115
3.3 Information...119
3.4 Shannon information in the mental space........125
3.5 Information in the mental space.......................128
3.6 Learning 132
3.7 Reduction and Inclusion 137
3.8 Complex adaptive systems.................................143
3.9 Causality 148
3.10 The mind and body problem 151
3.11 Mental causation 157
3.12 Cartesian dualism and fragmentation.............160
3.13 Libet's experiment 162
3.14 Why the mental space 165

IV: KNOWLEDGES, BRAINS AND COMPUTERS

4.1 Studying with Pythagoras 173
4.2 Knowing and Being 176
4.3 Knowledge and the quest for truth...................181
4.4 Leibniz dream 185
4.5 A deep mathematical crisis................................191
4.6 Gödel and Hilbert...193
4.7 Algorithms and thinking 198
4.8 The decision problem..201
4.9 Undecidable problems are everywhere............205
4.10 Oracle Machine and natural worlds 209
4.11 About computer simulations 211
4.12 Extending the explanatory landscape.............213

V: FROM INVISIBLE WORLD TO SCIENTIFIC PARADIGMS

5.1 Mental spaces and invisible worlds 219
5.2 The language of Gods 222
5.3 Physical laws...227
5.4 Subjective choices in the scientific method.....234
5.5 Sciences and mathematics.................................239
5.6 The Circle of Vienna 245
5.7 Paradigm shifts..251
5.8 The Laplace deterministic Universe 256
5.9 Physicalism 259
5.10 Computationalism is a bad brain model.........263
5.11 Why is computationalism wrong?..................267

VI: UNKNOWABLES

6.1 Undefinable words, antinomy 271
6.2 Truth and belief..276
6.3 Existence in physics...283
6.4 Causality and special relativity 286
6.5 Understanding understanding 290
6.6 Intuition..294
6.7 Free will 303
6.8 Time 305
6.9 Limits to scientific knowledge.........................309
6.10 Self-reference and blind spots........................311
6.11 Intelligence, Artificial Intelligence 318

VII: WE HUMANS

7.1 The imitation fallacy 323

CONTENTS

7.2 Fighting emptiness by learning 329
7.3 Conclusions 334

Preface

> "Science is done by humans. This obvious fact is easily forgotten; it is perhaps useful to remind it." Werner Heisenberg

This essay is about how we produce "realities." Scientists think of the world under the "External Reality Paradigm or Postulate" (ERP), whose familiar name is "realism." ERP asserts that our mental representations reflect what "is" out there. One can correct eventual "reading mistakes" by empirical techniques.

Brain-centrism, on the contrary, asserts that our mental representations are not what "is" out there; we perceive our reaction to the external world. We can correct reading errors, but cannot change the structure of the mental space that identifies and understands. One cannot "correct" the fact that we all are human observers. No one has seen the universe neither through non-human eyes, or with a non-human brain.

Recent progress in cognitive neuroscience has forced us to question ERP and distinguish "what is out there" from how our mental space illustrates it.

Brain-centrism argues that our so-called objective "third-party" statements are "first-party" human stories structured according to precise rules. These rules, known as the scientific method, should make sure that knowledge got by respecting them is the best one can expect.

ERP proposes an improbable situation: an isomorphism between "out there" and our representations of "out there." We shall explain all along with this book why this extreme case is so unlikely.

In this first volume, we shall examine the "mental space," an abstract entity generated by the brain's activity. The mental space is the place where we live, feel, believe, desire, fear, and think. Its components are mental representations. We shall examine its relationship with the brain, and its connections with "out there." We shall discuss how language developed and modeled our realities.

Many properties attributed to the universe are, in the brain-centric approach, properties of the mental space. They derive from the brain's physiology. Causes, that ERP refers to as belonging to the universe, are sometimes brain productions.

Only a non-human entity could confront us with a different view of "out there." A view not "readjusted" by our mental space. We shall examine what attributes an extraterrestrial mental space could have. We shall also examine the computer's mental space.

Special and General Relativity, diverging with Newton's approach, has brought in some physical characteristics of the observer, his speed and acceleration. Brain-centrism goes one step further, recognizing some aspects of the mental space in our explanatory landscape.

The controversy between ERP and variations of brain-centrism has been ongoing. Different human cultures, including the Chinese and the ancient Greeks, have considered forms of brain-centrism.

Important figures such as Leibniz, Kant, Schopenhauer, and Einstein, Bohr, Schrödinger, to name a few, could be precursors.

Brain-centrism is significant in science, but it also opens new avenues in social sciences, the economy, and most human endeavors.

Brain-centric will comprise two parts. I entitled this first part, "The Mental Space." The title of the second part will be "Science and Knowledge viewed from the Mental Space perspective."

The first three chapters examine the functioning of the mental space. Chapter 4 is dedicated to knowledge and the quest for truth. Chapter 5 analyzes how and why the mental space generated invisible worlds, mathematics, and science. Chapter six concerns the limits of the mental space and primitive undefinable words. Chapter 7 is a conclusion. A few paragraphs may be harder to read, please just skip them. The book depicts the concept of brain-centrism under various angles, chooses the ones who are the most interesting for you.

*

History, why this book

The quest for a consistent and comprehensive global worldview has occupied most of my thinking life. Was it an obsession, was it my way to rebel against an established order, or against the prevailing thinking paradigms? I cannot know. I realize I always preferred non-conventional thinking and alternative approaches to connect the dots. All our dots regroup in the human brain. One may study a remote galaxy, celebrate an anniversary, play tennis, or struggle to solve a complex mathematical question; all take place in our brain. The brain is the focus of the mystery. And the mystery is the source of any quest.

Let me explain how I came to the brain-centric perspective. The first significant step was when I studied Kurt Gödel a few years after my Ph.D. Logic was not my specialty. However, the incompleteness theorems resonated for me far beyond their mathematical horizon. It was as if I knew, since very young, that our explanations of reality did not present us with what was going on out there. I had read Plato, but with Gödel, we have a mathematical theorem, not a simple allegory. The weirdness of quantum physics and its total dependence on mathematics had increased my suspicions. Could He have conceived something so bizarre? The equations work fine, but there must be another way to interpret them! Is there a "superior" necessity for things to be like that?

The next big step, reinforcing my doubts, was reading, some forty years ago, the work of Alfred Korzybski, proclaiming that "the map is not the territory." I took time to digest it. I was not sure if Korzybski's idea was evidence or a revolutionary vision.

I tried relating incompleteness, Korzybski, and quantum measurement. That is when the writing of this book started.

Twenty years later, after many reading books and having innumerable discussions, I was still trying to imagine a more satisfactory view. At the Swiss polytechnic in Lausanne, where I was teaching, I met Professor Henry Markram, a distinguished neuroscientist who had just arrived from Israel. He wanted to set up a crazy project. He planned to simulate the brain on a digital computer. And he convinced me it was possible in the next ten years. I became associated with the project with some other friends and professors from EPFL.

The Blue Brain Project allowed me to confront my views on reality with dozens of scientists worldwide. With one exception, none of the thinkers I met gave me the impression that their ideas would help me in my quest to connect the dots. They were specialists; I was lucky enough to be an amateur. The "one" exception was a Brazilian neuroscientist, a professor at Duke University, whose openness, humanistic perspective, and extensive culture impressed me further and further at every encounter. Although Miguel Nicolelis is a renowned neuroscientist, his interests went well beyond his academic specialty. My ideas resonated with his, and we became very close friends, discussing for years our favorite subjects on a quasi-daily basis. He taught me most of what I have learned about the brain.

I abandoned Platonism, and neuroscience convinced me that perception could not tell us all about the fabric of "out there." However, I could not surmount the key issue of anyone that was educated in ERP:

What I see "is" out there, I can touch it, smell it, kick it. From what I see and think, I can build instruments, and these instruments work as I had proposed.

The relation between my representations and out there seemed to be direct and truthful: I see what "is" there. We all see the "same thing," we can study and describe it to one another.

We can predict the movement of the planets and verify our predictions. We can even collide particles and expect the result.

Realism, physicalism, materialism, and even computationalism must be true! But then consciousness and free will must be some illusion.

Gödel helped me again. The high-power digital computer that IBM had provided to Blue Brain was using a language like any digital computer. Step-by-step instructions. Gödel's incompleteness suggested to me this system would never simulate the human mental space. Computers cannot understand. Computationalism must be a false hypothesis.

In 2011 (having left the Blue Brain Project), I wrote the first essay on the topic, asserting that a mechanism was something different from an organism. We build mechanisms, organisms evolve. I explained in this book that a simulation of the brain on a digital computer had a minimal meaning and would teach us very little. I argued the brain must have an analog component interacting with its digital neuronal network.

Miguel Nicolelis, although he is the founder of BMI, brain-machine interface, had the same opinion and the same ideas. He thought computationalism was a false assumption for very different reasons connected to how populations of neurons handle sensory information. We associated our efforts to publish in 2015 a monograph: "The Relativistic brain, how it works, and why it cannot be simulated by a Turing machine."

In this monograph, we extended the idea of information by considering the information that was non-reachable by a formal system, as Gödel had described. We called that information Gödelian information. Contrary to Shannon's information, Gödelian information is not symbolic and digital. It is rather physical, analog, and integrated. Shannon information cannot be causal effective because it is "subtract independent," Gödelian information is because it is physical and subtract dependent.

These last 15 years, most discussions I had about reality have taken place with Miguel Nicolelis, who became over time more than a friend, a soul partner, or an intellectual brother. Confronting our two views, the one of a distinguished Brazilian pioneering neuroscientist and the one of a mathematician engaged in foundational questions, has been the most fertile adventure of my whole intellectual life. I wish it could continue until my last minute.

What convinced me that brain-centrism was the most compelling perspective was an amazing personal experience I went through in December 2015. After serious heart surgery, they held me in an artificial coma for ten days. For ten days, I lived in another world made of "narcotic dreams" just as convincing as what we call reality. The experience was so fascinating that, for years, I could not figure out what was real in these dreams and what was pure brain inventions. I wrote a book in 2016 to narrate these amazing narcotic memories.

Brain-centric is thus the result of a lifetime of questioning to understand the nature of human reality.

Miguel has published a book (2020) entitled "The True Creator of everything: How the Human Brain Shaped the Universe as We Know It." The theme is similar to "Brain-centric" but The True Creator is enriched by Miguel's vast culture in neuroscience and human history. I can only recommend reading it; I had the honor of reviewing the preprints, and it is breathtaking.

*

Notes to the reader

I shall often use the word "out there" instead of "reality" or "universe." "Out there" bears fewer inferences and ERP references.

I used "mental space" and not psyche or mind, which bears their references. Studying the mental space differs from psychology.

The individuations we will use, such as the natural world, and the verbal world have no correspondence in the psychological language. The goal of psychology is healing diseases. It is far from the purpose of brain-centrism.

I did my best for the book to present a linear development, where I define a concept before I use it. I failed miserably. I use many ideas and only develop them later. Skip any paragraph of lesser interest for you. However, I added a reference to the paragraph where I expose the concept. I also exposed many ideas several times at different angles according to the context, hoping to make things clearer.

In many aspects, this is not a science, semantics, or a philosophy book. Many of the exposed approaches are not consensually accepted science, which is ERP-based. I hope to start discussions and contradictions.

I have included several historical narratives in this book, hoping to make clearer the long cultural route towards brain-centrism.

For my readers who want to have a faster overview of brain-centric, I marked with a star some key paragraphs.

Finally, one can object I have written this book before all the ideas are sufficiently mature. I know my only excuse is age…

PERCEPTION

I: PERCEPTION

"The laws of physics are believed to be at least approximately true, although they are not logically necessary; the evidence for them is empirical. All empirical evidence consists, in the last analysis, of perceptions; thus the world of physics must be, in some sense, continuous with the world of our perceptions, since it is the latter which supplies the evidence for the laws of physics." Bertrand Russell, 1927.

1.1 Survival

The human brain has not developed through millions of years to maximize our intellectual reasoning skills or to perform mathematics. That was not the main orientation of evolution.

The central nervous system (CNS) formed to regulate and facilitate our movements, to extend our survival skills. Controlling the inner motor system was the function of primitive mammals and human CNS.

It was not decisive to bring about a perfect representation of the "external reality." Realism was not the name of the game. Much more critical was to generate representations that gave the best chances of survival. Even if, from a third-party perspective, the picture presented was statistically wrong.

*

Let us place ourselves about 50,000 years ago.

Imagine that beautiful sunny evening in the African savanna. A man has removed from his group, intrigued by an unusual odor. The senses of this primitive Homo sapiens are as always on alert. His eyes scrutinize the landscape, and his ears interpret every sound. His sensitivity to a range of smells also contributes to his global awareness. His well-developed primate brain is anticipating what could happen and seeks sensory evidence of his expectations. Most of this activity is analog and instinctive; consciousness would slow the process. Memorized visual, olfactive, haptic, and auditory maps, at various resolution levels, meet with sensory signals, adjust, confirm, and readjust, generating a high level of brain activity.

Our man is experienced. He knows how to account for a vast number of signals including the wind direction, the humidity, and the temperature, although he has no names for them. He can also feel and know where his fellows Homo sapiens are. His "amygdala" assesses the environment for danger, ready to conjure the fear he might need to respond to any threat. His state of alertness is maximal, controlled by his reticular activating system (RAS) and his neurohormones. Any activity in the sensory pathways leading to his cortex will activate the RAS, which will facilitate the cortex, which allows the processing of the combined sensory input.

A sudden gust of wind blowing on the high grasses produces a slight, unusual vibration. This vibration triggers an electrical signal. It starts in the ear, which confirms the details of a representation proposed by the memory's anticipatory system.

This analog match, regarding a minor detail, triggers a feedback loop between the RAS and the cortex. Neurotransmitters get liberated and activate higher levels of alertness and synchronization.

The anticipatory representation proposes as an "interpretation" of the gust of wind: a lion crouching in the grass.

Natural selection has taught him not to hesitate and to accept this map as "real" and rush away in the direction opposite the signal.

Believing a grass tremor is the sign of a lion is effective for survival, although it is often false.

No similar abusive interpretation applies to his internal body sensors. The information pathways are different, and the coordination of his muscles is perfect. Like most mammals, he will not choose a straight line for his escape; it is much too predictable to be safe.

Nature has invented the fastest way to generate safe (although wrong) representations of external reality. This quickest way is by generating expectations. To analyze, compare, measure, using various thinking procedures would slow a decision. On top of the shortlist of expectations, his mental space selected the ones that signal danger.

The brain has evolved to maximize survival, and not for truth or "realism."

* *

1.2 Perception

Brains rely on anticipation and prediction mechanisms to align our perceptions with our expectations. This involves two separate groups of neurons. One population of neurons encodes the sensory data coming in from "out there," and another group represents the memorized expectations.

Each confrontation of the encoded sensory data with pre-existing representations contributes to refine our expectation list and its memorization. Through these continuous interactions, the brain builds new neuronal networks. From them emerges our multidimensional inner "reality." All our sensory perceptions and their associated feelings combine at various resolution levels to improve these constructs. These multidimensional representations, and their interactions, constitute the *"mental space,"* as opposed to the underlying brain who would be the *"physical space."*

Large-scale connectivity between brain regions makes it possible to privilege some expectations according to the context and personal history. The brain is continuously refining, combining, and adapting its expectations. The more often we activate a connection, the more it builds up, and more we will use it. Prior knowledge can, therefore, reinforce expected signals and reduce the influence of less-awaited signs. Perception is thus a selective process, started in the brain, where incoming messages by confirming an expectation allow it to gain a higher priority. Anticipation is a mechanism used by all our senses and our feelings, who represent our physical emotions.

If sensory information contradicts your expectations, you feel surprised and alerted. Your brain releases adrenaline, and for some milliseconds seems to stop working, as if paralyzed. It is trying to integrate, as fast as possible, the latter information incoming from the senses, adapt or call up a better matching prior knowledge to determine what to do. We know this behavior as *"episodic simulation."* Expectations are at work in every facet of our life, and we cannot avoid thinking about every future situation. That is what our mental space does in the background all the time.

Expecting is one of the unstoppable mental mechanisms.

Anticipation selects what information our senses will prioritize in their probing. The selection process allows us, for instance, to single out and follow a conversation in a noisy party gathering, or to spot out the spider that you fear on the wall. The permanent confrontation between our inner models and our sensory information determines our views about the external world. It develops the individual we are and the reality we live in; it shapes our mental space.

We expect somatosensory stimuli, such as heat, cold, touch, and pain, before being confirmed.

When you put a step forward, you "expect" the ground to be robust under your foot. Have you tried climbing a non-functioning escalator? At the moment you step off your brain behaves as if the escalator was working, and you risk losing your balance. Speaking to someone, you expect him to listen and react. When you receive the meal you ordered, you expect its smell and its taste.

We handle perception differently if it is conscious or unconscious. Unconscious perceptions involve representations that analyze colors, sounds, shapes, faces. The same object shown consciously involves a much larger variety of representations, activating regions beyond the sensory cortex and recruiting a more significant neuronal synchronization basis.

A study was published in 2017 by the neuroscientist Peter Kok, Pim Mostert, and Floris de Lange. These scientists show that, when we expect an event to occur, one can detect activity in visual cortex well before the event out there happens. This activity provides a considerable speed advantage, allowing us to adapt our behaviors. When the expectation which occurred before the fact is wrong, the goalkeeper hurls himself in the opposite direction!

As feelings such as fear, pleasure, or desire preceded the development of other representations, we associate them with most expectations. Their frequent and ancient usage raises them high on the list of what the brain expects. It takes hard personal work to control them.

Because of its strong emotional roots, we consider music must have been an extremely ancient mental activity.

We love themes where we can expect the succession of notes. When the sequence is too simple, too expectable, the music feels dull. Successful music pieces subtly combine satisfied expectations and moments of surprise. We would feel a total surprise as a *"random succession"* of notes and not consider it as music but as noise. It's impossible to expect! Noise is not singable nor learnable. To memorize we need order! (2.6) Music must follow some expectable patterns for our brain to recognize it, to remember it, and ultimately to "enjoy it."

A proper balance between routines and innovation is a condition for enjoying any activity. We feel repetitive tasks as boring. A rapid succession of unforeseen events can destabilize us. It undermines any of our predictions.

Expectations and representations are not physical, although our physical brain activity generates them. They are mental/informational *experienced* events. They belong to the mental space, and no third-party viewer can observe them. The best third party could do is measure the underlying neuronal activity. The mental space shapes what we experience. We can voluntarily associate, combine, adapt experiences to our needs, memorize them and generate new imaginary ones. However, most of the time, our expectations remain unconscious.

Because expectations are mental, they give our perceptual systems incredible flexibility and the capability to spot out similarities and create new representations.

Without the "motor" of analogies, no thinking, creating or imagining would be possible. Analogies build links between representations. They build and organize the mental space in a dynamic and continuous process. New connections develop and older ones disappear in the underlying physical space.

For instance, by looking at the clouds, one can see a face or a dog that someone else would not notice. By staring at the stars, we can draw familiar objects, the constellations, if we are told what we should see. Manipulating representations allows us to "imagine" and develop new ones extending far beyond what we perceive. By doing so, we generate new mental states. Artists are used to this extraordinary inflation of their mental space. Applied scientists and judges, on the contrary, learn to avoid any mental extension during their professional life. We train them to be "realist," and stick to "facts." This ambivalence is because the mental space has two facets. (1.7)

Mastering what we expect is mastering our perceptions. We can act on expectations, changing their priority and organization. We first need to become conscious of what we are expecting. Humans are perhaps the only animals who can reorganize their expectations according to conscious wills. This capacity to interfere with spontaneous reactions explains the immense diversity of knowledge and worldview. Even more than biological diversity, mental diversity is responsible for the incredible and unlikely success of our species.

The mental space can thus act on itself. It can represent itself as one of its components. This group of representations is our self-image or self-schema. It is central to any representation network and interferes with any perception or thinking process. Because of self-image, the mental space has its point of view, like an origin in a reference frame. A positive self-image is vital because every perception reinforces it. From the outside, third parties perceive it as our personality.

No finite narrative can encapsulate the mental space. There will be a part of it who escapes any theory or description.

The mental space can observe itself from the outside of any description, and change thus falsifying it. We are here pointing at the very source of free will, and the reason no precise scientific theory of the mental space can describe it.

Representations and expectations are building blocks of our multidimensional hyper-connected mental space. Expectations are a driving force behind its activity. In this space, we experience all we can know about our lives and our universe. One considers that the brain gives us access, or generates our mental space. This is because the mental space vanishes when the brain dies, at least from a third-party perspective. If we can observe and measure the physical space, we only know the mental space from the inside as first-party knowledge. If the physical spaces of various individuals look very similar, their mental spaces can be unimaginably diverse.

Brain-centrism considers that all our knowledge is about our representations. These cannot reflect the nature of what is out there. ERP is the assumption that we perceive what is "out there." Perception is a direct isomorphic image. ERP has developed in connection with vision because of its predominance over all other senses[1].

If one considers taste or smell, for instance, ERP appears much less consistent.

We cannot find sweetness in the sugar by analyzing atoms or molecules, it needs a mental space to taste it.

We consider odors as a minor factor, but they are critical for us. Losing smell, called anosmia, disrupts virtually every facet of our lives and transforms our inner reality. These disruptions can concern:

our hygiene, a loss of sexual intimacy, a loss of appetite and weight, a loss of memory, modifications to our emotional balance, depression, anxiety, and the breakdown of personal relationships.

How did science come to adopt perception by expectation?

Hermann Weyl (1885–1955) was one of the greatest mathematicians and physicists of the last century. Here is how he described the unexplainable mysteries that raise a direct perception process.

"The processes on the retina produce excitations, which are conducted to the brain in the optic nerves, as electric currents. Even here, we are still in the real sphere. But between the physical processes which are released in the terminal organ of the nervous conductors in the central brain and the image which thereupon appears to the perceiving subject, there gapes a hiatus, an abyss which no realistic conception of the world can span. It is the transition from the world of being, to the world of appearing image or of consciousness."

The so-called Binding Problem shows up in the computational model of the brain. Different sensory information reports an

---

[1] the processing of visual information seems to dominate the processing of information from other senses in humans. Other animals gain most of their information through their sense of smell.

event using distinct brain circuits to encode that information. However, the brain generates a coherent, precise perception out of these non-synchronized data. We accurately perceive the consolidation of different sensory messages, although these signals reach the brain unsynchronized. The speed of the signal along the distinct pathways is different. In the computational model of the brain, this synchronization of the various auditory, visual, and haptic stimuli is unexplainable.

The binding problem appears because we consider we have created the representation "a posteriori" (or after the stimulus). In the anticipatory mechanism, the brain generates its expectations of what is being seen as "a priori." Incoming signals interfere with the "a priori" expectations until a resulting ultimate image emerges. In this model, there is no "Binding Problem." Everything is "pre-bound" by the expectation and confirmed by the various incoming signals.

In the 19th century, Hermann von Helmholtz (1821–1894), known for his theory of vision, proposed the idea of *"unconscious inferences."* For him, the observer uses unconsciously tacit previous knowledge, which, combined with the sensory input, provides an "accurate" representation. He was one of the first modern scientists to express doubts about ERP.

The philosopher John Friedrich Herbart (1776–1841) that inspired Bernhard Riemann is a forerunner of brain-centrism. As a successor of Immanuel Kant, he was a pioneer in the study of perception and learning. His great concern was the relation between mental representations and sensations. For Herbart, vision and touch developed by combining the *"production of space"* in our minds. He describes visual images as mental hypotheses, adapting to the sensory inputs. By considering space as a mind production, his ideas inspired Riemann's discovery of non-Euclidian geometries. Riemann's discovery is the symbol of the inspiration brain-centrism could generate by overcoming the rigidity of the ERP dictate.

Alfred North Whitehead (1861–1947) was an important mathematician and philosopher. He co-authored with Bertrand Russell of the three volumes of the *"Principia Mathematica."* The prevailing perception theory embarrassed Whitehead. In his book, *"The Concept of Nature,"* he writes:

"What we see depends on light entering the eye. Furthermore, we do not even perceive what enters the eye. The things transmitted are waves or—as Newton thought—minute particles, and the things seen are colors. ... There is now reigning in philosophy and in science an apathetic acquiescence in the conclusion that no coherent account can be given of nature as it is disclosed to us in sense-awareness, without dragging in its relation to mind."

In the 1970s, the British psychologist Richard Gregory (1923–2010), founder of the Department of Machine Intelligence and Perception at Edinburg University, suggested, *"perception as a hypothesis."* He compares human perception to the way the scientific method works, conjecturing followed by verifying.

Stephen Hawking and Leonard Mlodinow introduced in their book "The Grand Design" (2010), the *"model-dependent realism."* Our brains interpret (using memorized expectations) the input from our sensory organs and make a model of the outside world. We then organize these models in mental concepts that become the only "reality" we can know.

In his book *"Beyond Boundaries"* (2011), my very dear friend and brother Miguel Nicolelis assert:

"Most information about the world and our body comes to the brain because of exploratory actions initiated by the brain itself. Perception is an active process that starts inside our head..."

He was the first to speak about the brain's own point of view, expressing doubts on ERP. Miguel confirms these doubts in his recent book: "The True Creator of Everything" (2020). Miguel has established the expectation mechanism as the dominant position.

Scientists who initially proposed an *"expectation mechanism"* could not realize how deep and far-reaching this proposal would end up being. They could not have imagined the fantastic consequences that it would have for our understanding of reality. The expectation mechanism will play a central role in the brain-centric idea, we consider it as the neuroscience equivalent to the Copernicus Heliocentrism.

Non-observable components make a large part of our reality; therefore, expectations will play a critical role in our worldview.

* *

1.3 Expectations

When you meet a stranger for the first time, you are cautious because you do not know what might happen. He registers in a category of your mental space as "stranger," and strangers are unknown and may be dangerous. You might feel cautious and try to find out details that support this impression.

If a friend has introduced you, it makes things more comfortable because you know a bit more about what you can expect. Over time, you get to learn about him in many situations. As you update your expectations, you become less defensive. The stranger is the same person, but your representation has developed.

In our social relations, in business, in negotiations, and in friendship, we need to show predictable behaviors to make our partners feel comfortable. That is how we build a reputation. People know what to expect from us (or our brand).

Reading this text, you only read the first letters of each word or even of each sentence. Your brain's expectations build the rest of the word or the sentence.

When you hear or read a sentence, understanding it results from associating/connecting it to a multitude of prior representations. This prior knowledge includes multiple dimensions: the social context, feelings, memorized narratives, intentions, relationship with the speaker, etc. A large part of your emotional, physical, and cognitive history takes part in your understanding.

When a child learns how to read, the process is slow and painful; it requires the continuous intervention of consciousness. The child has to read every letter to build up an expectation catalog. Adults already have patterns for most words and sentences; therefore, a simple hint is sufficient for their brain to decide what comes next.

When listening to another person, we put ourselves in his situation. This involves simulating his state of mind and trying to make sense of it. This amazing capacity of the mental space, named empathy, involves imitation and analogy. It implicates specialized groups of neurons, called mirror neurons, distributed in various brain regions. Mirror neurons can synchronize with other mental spaces. They are not only helpful to understand others, they also enrich the states of our mental space and its expectations through learning. Speaking with an old-time friend, you can guess his words, a feeling that will enhance your receptivity inhabits you. The expectation mechanism plays its role in empathy. When we perceive a facial expression, for instance, it triggers expectations about his feelings and his mental state.

Expectations are sometimes misleading. One can spot many pictures on the internet, created to be interpreted in various manners, depending on what the viewer expects. You can either see an older woman or a young one, either a rabbit or a bowl of vegetables. It depends on your expectations. These pictures also show us how complex it is to change one's expectations. Once you identify the old lady, it is rather difficult to switch and see the young one. We recognize what we expect to see and have difficulties seeing what we do not expect.

I suggest you examine the picture (Figure 1.3.1) before you turn the book upside down.

Once you activated an expectation, it behaves like a habit; it pops up in every similar situation, and it is challenging to change.

Our brain seeks confirmations; it does not seek falsifications. We call this process *confirmation bias*; it plays a significant role in stabilizing our worldview.

Belief echoes are also common perception and memory biases. They happen when you have listened to a narrative and determined it is incorrect. Later, if you hear it again, you might give it some credibility; you have remembered knowing the story but have forgotten your initial judgment. That is often a strategy used in politics to create rumors. "Tell your negative story, even if people doubt it, something will remain."

We feel other people's feelings and wishes and react when they are expecting something from us. Children should learn young how to say no and not respond to all expectations. We do not limit empathy to our species; our sensitivity extends to other mammals' feelings and moods.

The *focusing effect* is another common bias linked to the expectation mechanism. It occurs when someone places too much emphasis on a picked detail rather than considering the "larger picture." Because of the emotional importance it has for him, he sees this detail as essential, whereas for other people it would not be relevant. The excessive focusing effect is due to how feelings relate to expectations.

Focusing effects also produces *selective perception*. The first time you get a dog, you notice many people are dog owners. The first time you buy a red car, you will cross a lot of red cars. By imagining our future, we use the focusing effect to help us reach our goals, putting the expectation mechanism at our service. Through this effect, fears, desires, and values govern our lives without us even noticing it.

We estimate that 80% of plane crashes are because of pilots' errors. The error stems from physiological and psychological human limitations. Most pilot errors are because of a wrong interpretation, of an alarm signal, of the scenery, of an instruction from the tower... Expectations of the pilot play an important role in generating these errors.

Brain-centrism asserts that we have no access to "true reality," we cannot even know if such a "thing" exists or what "existing" could mean for such an entity. What our mental space presents us is the only reality. Our fate is thus to build the best possible mental models. Determining what we should mean by best possible is the crucial question. For our savanna man, the "best model" relates to survival, but since that time, the environment in which we must survive has changed.

We can detect and correct perception biases, like in the examples we analyzed in this paragraph, by comparing with third parties. But biases that would result from the structure of the human mental space are undetectable and unavoidable. Only a non-human could act as a third party.

**

1.4 Filling in

"*Filling in*" is the second crucial mental space unconscious mechanism we shall consider. It is also a non-stoppable driving motor of the mental space activity. While building mental representations, some sensory data will be missing. The brain "fills in" the missing elements by interpolation. It "guesses" what should be there according to the situation, its history, and its expectations. Filling-in is a perceptual phenomenon. The brain adds information in the perceptual field, although no signal has reached it.

Our human eye has a physiological blind spot. This is where the visual field that has no light-detecting photoreceptor cells. At this place, the optic nerve passes through the optic disc. No vision is possible there. No photons arrive in that region. However there is no "hole" in our visual field. The brain has filled in.

The transmission speed of the nerve fibers is about one meter per second. When you blink, you blackout some 80 milliseconds which your brain fills in, stitching the gap, without you even noticing. Enjoying a movie or watching television involves filling in.

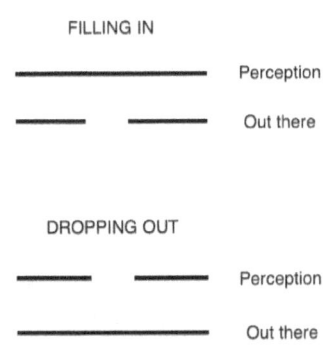

Figure 1.4.1: Filling in and dropping out

The same thing happens when you listen to an indistinct conversation. A prestigious bottle creates an expectation on the taste of the wine, one often fills in with the expected taste.

Filling in happens at all levels in the brain—when we have missing data, the brain builds the data, invents them by using analogy, expectations or interpolations.

The reverse process, sometimes called *dropping out,* can also happen. When something is not expected to be there, we sometimes do not see it! In a famous experiment, the spectators are presented a clip of a ballroom with people dancing. They are then asked if they noticed something unusual. Most people notice nothing, although there is a man disguised as a gorilla who crosses the scene. The mental space wipes him out because he does not fit in.

The brain initiates perception by probing expectations. It then confronts them with multiple sensory input data. It organizes these data, synchronizing and combining them in a coherent picture by filling in the missing parts and dropping out the incompatible parts. This happened, for instance, when observing the image of the young lady above.

The fill-in mechanism is a cause of false and catastrophic representations. We attribute car and even plane crashes to fill-ins, called human errors. Training can teach us to avoid many of these human errors. Over-training can also favor their production. Having exercised so many cases, our mental space might find a case close enough to fit the incoming data, but wrong.

Fill-in has critical survival advantages: it is fast, it allows the brain not to get stuck by a missing piece of data to decide. An ill-advised choice is preferable to no decision at all, filling-in developed as perceptual guessing, educated by trial and error. Often errors could be fatal. Once language appeared, the functions of filling-in increased, generating a vast variety of creative narratives modeling our inner worldview.

Filling-in is so tied up in the way our mental space gains knowledge that one cannot dismiss it. Human civilizations have been using fill-ins without noticing mistakes for centuries. Filling-in is part of how our brain and its mental space work.

Most of the time, shaped by millennia of natural selection, our expectations "fit" well enough with what is "out there." They allow us to "navigate" the world. However, occasionally they do not! With some understanding of how expectations work, one can create pictures, movies, or virtual realities that "cheat" our interpretations and generate "illusions."

A typical example is a TV set. What you see if you're in front of the screen is a James Bond film. A technician "knows" that the electronics switches on and off in a precise order, a series of pixels.

Describing the scene, you and the technician would deliver two distinct languages. You would speak about James Bond diving from a helicopter (the first-person perspective); the technician would explain how the intensity of 5 million LEDs varied over the last minute. (Third-person perspective.) You are describing what you perceive and feel; he is saying what he has learned about how to produce TV images.

It makes no sense to suggest that one description is more "real" than the other. The third party description is more appropriate to repair a TV set. The first person one, to have an entertaining moment. You and the technician are not "using" the same mental information sources. Both are useful and both appear in mental spaces.

* *

1.5 Individuation

Detecting movement has been one of the most critical functions of the visual system. Any movement detected out there could mean danger or food and relates to survival. We infer motion perception from the changing pattern of light in the retinal image. One detects differences in size, texture, contrast, and other characteristics. The Middle Temporal (MT) area of the primate visual cortex receives and processes the visual signal. When we detect unexpected changes, the MT triggers the RAS and facilitates the cortex and immediate awareness. Movie producers know how to control our attention through movement, sound, and image. Preys have learned to immobilize when they perceive danger. Predators could notice a movement of escape.

Luminance, contrast, change, and other factors allow the mental space to differentiate one object from its environment and other objects. Touch, vision, and sometimes audition collaborate to confirm the boundaries given to an object in a process called "individuation." We can test and sense differences because we can compare sensory inputs. It allows us to consider that something differs from its surrounding environment and forms an object, a unity per se. We can then identify this something, view it as isolated, generate representations, name it. This is a mental process, it is not something that happens out there.

The representation is independent of the object's movements or background changes. It's a network of individuated properties of the entity, including its subparts and all its attributes. Perceiving one of these properties will trigger an expectation of the other properties. That is why we imagine the lion by only seeing his tail. An odorous molecule is a sufficient attribute for the expectation mechanisms to propose a complete lion. Perception, according to psychologist Kurt Lewin, has three components, but we not only perceive an object and the environment, we also perceive the complex relations between the object and the environment. We integrate this third component into our representations and expectations, facilitating individuation when the object moves or changes.

The individuation process fragments our perception of "out there" into separated parts, building this complex multidimensional network of connected representations we called the mental space. Many interconnected representations thus represent the identity of an individuated entity. Once this identity stabilizes, the entity "exists" in our mental space. We are still far from understanding how this complex process works.

You notice a person advancing in your direction. Expectations you form concern a large variety of aspects that your brain has learned to predict. His physical look: "is it a man or a woman," the path he is following: "How far from me is he, how fast is he walking." But also his intentions "what is he holding in his hand, why does he seem agitated, what does he want from me?" According to the context, your "state of mind" will activate one set of expectations or another, then confirmed or denied by new sensory information flowing in.

Face recognition individuation and memorization are of the utmost importance for mammals. Looking at faces activates the fusiform gyrus, a region of the temporal lobe specialized in their identification. By stimulating the fusiform gyrus, scientists can alter your identification of a face. The patient will recognize someone else!

Learning a language will impose on the child the individuations and the relations built in the categories, the words, and the grammar of his language. Through the names given by the language, he will fragment, separate, individuate and reunite objects. The language will build a universe based on symbols he shares with other individuals of his group. It doesn't mean he shares the same feelings associated with the words.

Our mental space individuates not only external objects; it also needs to fragment physical emotions into different feelings and be able to associate them with names.

Although we experience feelings as "existing things," their individuation is not clear. Feelings don't have the precise and stable contours of external entities. They overlap, change, intensify, or vanish out of our lives. They are different for each of us. How to individuate and define things like intelligence, love, consciousness, friendship, honor, pain, infatuation, courage, respect, trust, shame, resentment, admiration, gratitude? Feelings lack the "objectivity" required for third-party individuation.

Human languages mirror this difficulty by allowing flexibility, metaphors, analogies, evocations, and poetry. Computer languages and mathematical definitions, on the contrary, must be rigorous to serve their purpose.

Society needs precise criteria and definitions, even for feelings. Defining a feeling with words can create difficult situations for children who are learning the definition and comparing it to what they feel.

If by reversing the definition procedure, instead of fitting the definition to the feeling, we impose that the feeling fits the definition, we might sacrifice some crucial aspects. This reversing is the source of what we call mechanization. We then constrain our mental space to behave according to a verbal definition, limiting it to exhibit what the rigorous definition allows.

Most children becoming teenagers confront this duality, a distance between what they feel and what "they should feel." Many have difficulties overcoming this discrepancy that can make them feel different or not adapted. It can take time for them to understand that the richness of our inner world is difficult to communicate in the apparent rigor of third-party perspectives.

Measurable individuations about human qualities can only concern behaviors, they never illustrate the richness of feelings.

Language is an incredibly powerful tool. However, to express its full power to describe our mental space, it cannot be rigorous, it must allow flexibility in the individuation of mental states; it must often be metaphorical. Music and poetry are more adapted to communicate feelings.

* *

1.6 Fragmentation

The great American theoretical physicist David Bohm (1917–1992), contributed in his original way to quantum physics. He expressed grave doubts about our methods of individuation of physical objects. He believed that our individuations and our fragmentation of "reality" required us to paint an inaccurate picture of what is there. His many conversations with the Indian philosopher Jiddu Krishnamurti from 1965 to the 1980s are remaining testaments of his thinking and the genuine passion of Bohm to meet truth. Some recordings of these discussions are available on the internet. When I was 20, I had the privilege to meet them in Gstaad, where Krishnamurti gave a summer seminar each year.

Bohm opposes "fragmentation" to "wholeness" and considers that our method of individuation relates to the limitations of our perception. Somehow we miss the point of understanding the universe:

"Things appear fragmented because we are looking too low; we fail to raise our sights to the level at which the fragmentation is only a part of a greater whole. As a result, we mistakenly see things as separate, as fragmented."

He also says in his book: "The Implicate Order: a New Approach to the Nature of Reality":

"Man sees and experience himself and his world as made up of components. Being guided by this view, man then acts in such a way as to try to break himself and the world up, so that all seems to correspond to his way of thinking. He thus obtains an apparent proof of the correctness of his fragmentary self-worldview, not noticing that it is he himself, acting according to his mode of thought, who has brought about the fragmentation that now seems to have an autonomous existence, independent of his will and of his desire."

Albert Einstein, who admired David Bohm, also expressed doubts about our fragmentation of reality. He wrote in a letter to his fellow physicist Robert Thornton in 1944:

"I fully agree with you about the significance and educational value of methodology, as well as history and philosophy of science. So many people today—and even professional scientists —seem to me like someone who has seen thousands of trees but has never seen a forest. A knowledge of the historical and philosophical background gives that independence from prejudices of his generation from which most scientists are suffering. This independence created by philosophical insight is the mark of distinction between a mere artisan or specialist and a real seeker after truth."

What worried Bohm is that fragmentations divide and separate things into concepts and words. "Out there," there is no such separation, but a "wholeness." The mental space generates fragmentation then uses it to discover fragmentation "out there" in a self-referral confirmation bias.

We then sometimes stumble into difficulties our fragmentation has generated, building a world of problems that have no existence "out there." We use the same approach to solve these problems, and we engage in unending self-referral loops. ERP presents individuations as "external" realities, being external they seem unchangeable, and we consider them as "facts." For brain-centrism, they are mental constructs, convenient for certain explanations, but in no way absolute.

Fragmentation establishes limits and frontiers create divisions that aren't in what we are trying to model. For instance, we separate between bodies and forces; we individuate a body like a planet and a force like gravity producing its movement. We have separated hardware from software. We separate mental and physical. Our material productions must carry this separation because the mental space produces them. An enormous difference separates an organism that developed as a whole, from a mechanism that was built as the sum of parts. Nature does not produce mechanisms, only humans do. (2.8)

ERP has many other complications: beyond fragmentation, we will examine them in this book. These difficulties make our models different from what they represent.

Brain-centrism argues that individuation and fragmentation are essential functions of the mental space's arsenal to gain knowledge. We cannot dismiss or correct them, as we can do it with optical illusions. Our individuations and fragmentation are part of the way we represent out there, they are surely not part of "out there" itself.

* *

1.7 Mental worlds

The prodigious complexity of the human mental space, the interdependence of its components, the subtlety and diversity of its activities will not fit in any summarized verbal description. We are describing one of the most elaborated complex adaptive systems in the universe. To understand it, our only possibility is to propose some simplified metaphor. The first simplification I suggest is to divide the mental space into two mental sectors: the natural world and the verbal world. This is the most central metaphor we will use to describe the mental space. The activity of these two worlds is very different. What we are and our reality will result from their unending interactions. (Figure 1.7.1.)

Two primary sources of information feed the mental space: the "internal" bodily information and the "external" sensory information. Both sources feed the natural world, the verbal world has no external connection. The first source is information associated with feelings from our physical emotions. The second is information triggered from "out there" by the first internal source through our senses. These two types of information combine and correlate to complete our natural world representations. (Figure 1.7.2.)

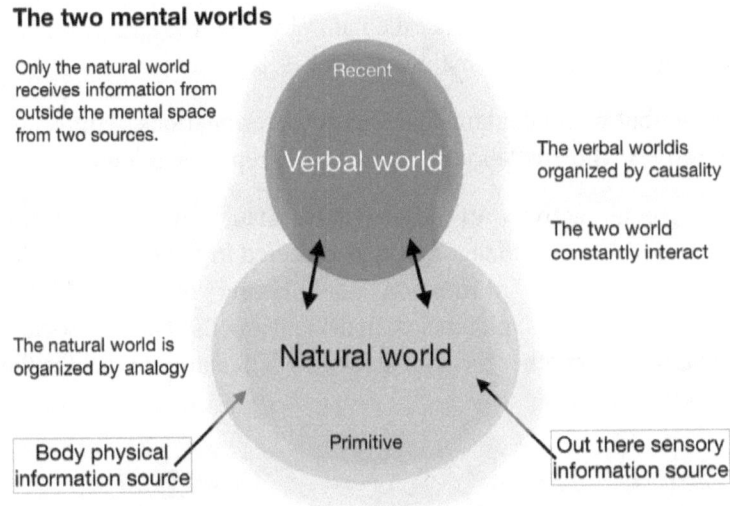

Figure 1.7.1: The mental space divided in two mental worlds

This is how the expectation mechanism materially works in the brain. We associate a representation with a feeling. The feeling "calls" the sensory representation and gives it its color: This association will be critical to understanding most of our high-level mental and brain functions including our sense of "meaning." (6.5)

The expectation mechanism always starts as a feeling that calls for, and associates with, a natural world representation. When we detect an anticipatory electrical signal in the neuronal system (3.12), the feeling has triggered it in the associated representation; it has called for.

Expectations are thus always "colored" by our emotional state. This emotional state that the mental space lives through analog feelings will determine what sensory representation we call for and how we "interpret" it. It is thus responsible for filling in and dropping out.

Internal analog information has thus a critical role in our natural world.

Our verbal world is "about" the natural world, the structures and categories of language mainly control its representations.

The verbal world contains our verbal representations, words, and narratives who are "about" natural world representations.

The mental activity of our savanna man, who did not use structured symbolic language, was centered in the natural world. Analogy organizes information and "thinking" in this world. It's a world of feelings, pictures, sounds, and smells. These associate by calling each other through analogies. The natural world is the

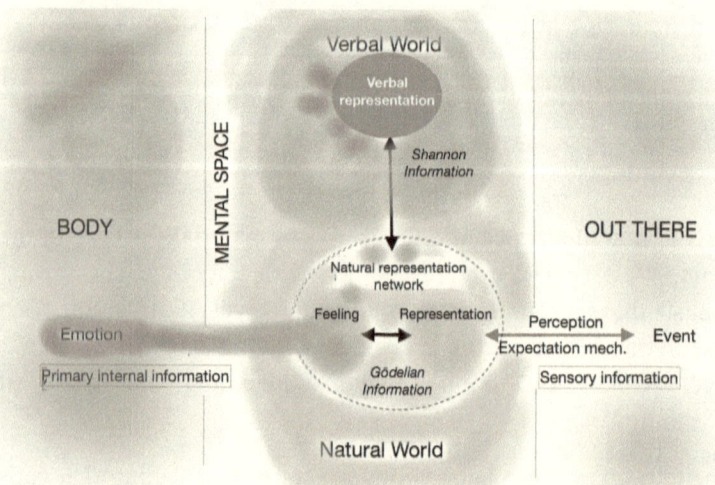

Figure 1.7.2: Two information sources

oldest mental world, the one in which our ancestors have spent their lives since our origins. It is also the world in which our children live until language becomes prevalent. This world of analogies, of sensations, of feelings, of imagination and of creativity will, however, remain crucial for us throughout our entire life.

Any representation immediately calls other representations they connect to. It will thus form a large network of representations associated with the initiating feeling to form the sensory perception. For any representation (R), our mental space will call up a network Net(R) of representations we call *"the logical context representation network."* The tail of the lion suffices to evoke the whole animal. Every element of Net(R) will have its feelings associated.[2] At the level of the brain neuron ensembles and synchronization of neuronal activity will build Net(R). Every time it is activated Net (R) will vary because the state of the brain and the sensory inputs will never be the same. We will often speak about a mental representation and rather mean Net(R) than (R).

Emotions do not use a representation system. They are physical. They were here millions of years before any symbolic representations and languages. We resent this world as truthful and natural. Here, things don't need to have a name to exist. Thinking in the natural world is a succession of images, smells, desires, impressions, and sensations organized by cascades of analogies. It's more like daydreaming or meditating. Spontaneity and creativity are in command. Time has nothing to do with clocks. Most representations of the natural world remain latent but active. They will influence our decisions and behaviors, our tastes, and expectations. They will determine our beliefs. Because they are unconscious, they are difficult to change, and words only are not sufficient. Through these basic emotional anchors, we give meaning to events and concepts. Without them things would remain mechanical, understanding itself would not

---

[2] The "material" brain version of what we described is the subject of our theory (with Miguel Nicolelis) of the brain exposed in our book *The Relativistic Brain*.

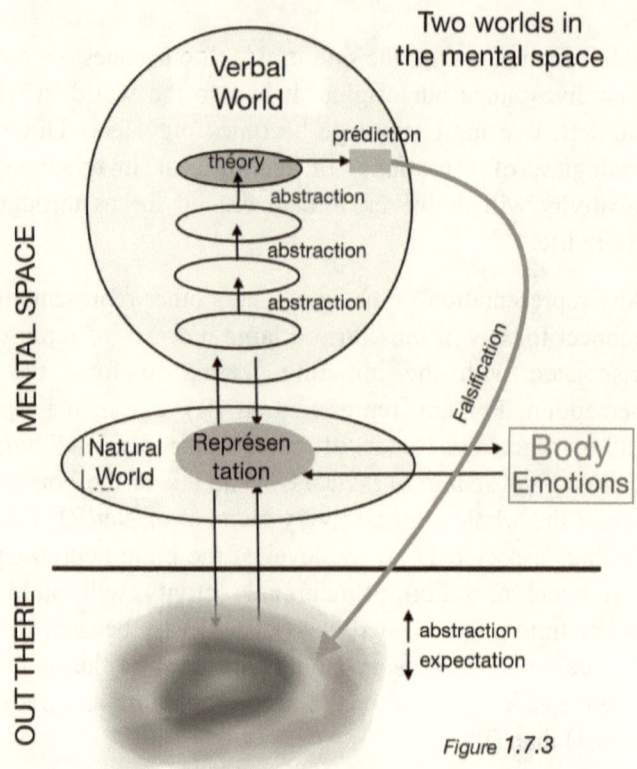

Figure 1.7.3

exist. In the natural world, there is no symbolic coding and no abstractions, there is no truth or untruth, things simply "are." The verbal world is "about" the natural world (Figure 1.7.3). It does not deal with "things" but with "meta-things," with things about the things. It is a level of abstraction higher, it's a meta-world. In the verbal world, thinking is telling stories. We already did that before the structured languages, but now language has changed the way we do it and we organize it. Causality has replaced analogy as an organizing force to relate events and concepts. Because we are dealing with symbols, we have the freedom to manipulate them as we wish. They can or not represent our natural world's feelings and representations. The major asset of the verbal world is that it allows easier and more precise communication between humans. They have now a common word to designate a subject even when it is not there or not material. Naming becomes a powerful tool to "give existence" to things. The categories and the structure of the language reinforce the individuation and fragmentation capacities of the verbal world.

This move from analog-to-digital represents the initial step of the digital revolution. For the first time on our planet, and perhaps in the universe, digitalization appeared in a brain. It was getting ready to change the planet, and perhaps a larger part of the universe. However, we should not forget who we are biological organisms, with a consciousness that strives to understand the universe and itself. A consciousness that forgets that the fundamental things happen behind its back at levels we cannot understand. Our brief history makes us short-sighted. Our first-person perspective puts us at the center of things. But it might well turn out our digital experience is only an ephemeral event, lasting a few thousand years. On this insignificant planet, something went wrong but nature quickly repaired it. We are not the player we are minor pieces on the chessboard.

* *

1.8 Two information sources

The feeling-dominated natural world is the oldest part of the mental space. Mammals millions of years before we discovered symbolic languages, already associated feelings expressing physical emotions to mental representations. The limbic brain, the structures that control emotions, motivation, and memories, is far older than the neocortex. Feelings, prolonging ancestral emotions, interfere with every aspect of our mental and social activities. They shape our representations, trigger our expectations, guide our thoughts and our intuitions, confirm our understanding, and determine our tastes and beliefs. Every mental representation, however, abstract it is, connects to feelings with bidirectional information. Human thinking, however abstract it is, will always be attached to feelings. Feelings drive the activity of the brain. That is why, as a child (and even later), it is so critical to learning how to balance emotions and thoughts. Associating positive feelings to our representations will determine our attitude when confronting any circumstance. The richness and variety of feelings will allow us to color our interpretations in a rainbow of amazing and creative tints or reduce them to limit categories, dominated by fear or greed.

Emotions are "physical," only their verbal representations are symbolic; that means their "information content" is integrated, "non-detachable" information. One cannot "detach" the information from the emotion itself. The emotion is the event, it is not "about" the event. Feeling something is living it, not commenting on it. We can say something that we don't think, we cannot feel something we do not feel. With my friend Miguel Nicolelis, we have called the non-symbolic, physical, analog information "lived" through emotions and feelings "Gödelian information," to distinguish it from the classical, detachable Shannon information. (3.4) One can say things "about" Gödelian information as we say something "about" this tree out there. But one cannot say Gödelian information, like one cannot say the tree itself. Gödelian information and trees are not words, they are not detachable from their physical expression, they don't belong to the verbal world. One says that Gödelian information is subtract-dependent.

This reminds us that the verbal world is not the totality, it is merely a superposed symbolic layer about the natural world. The information of this symbolic layer is subtract-independent Shannon information. The information of the natural world, be it regarding our body or any entity out there is unspeakable Gödelian information.

Another important difference between analog, physical Gödelian information and symbolic detachable Shannon information is that Shannon information does nothing. One says that it is not causal effective, it is only a succession of symbols that one could write as a series of zeros and ones. Such symbols can lie in a book: they are not active by themselves and mean nothing, as long as a mental space does not interpret them. Gödelian information is causal effective, it acts on the physical because it is itself physical. We can express this with a metaphor by saying that.

Planets do not know Newton's law. However, they orbit the sun.

Out there any information is analog, physical, and embedded, it is not detachable. To detach it, a mental space fragments, and separates information from matter or movement. That fragmentation only occurs mentally. It needs a brain. Having generated such a brain must have been a revolution for the universe. It took him thirteen billion years to get there (in our fragmented vision).

* *

1.9 Are expectations universal?

Do mosquitoes or extraterrestrials also use expectation and fill-in mechanisms to represent, "out there"? Do they individuate and fragment as we do? Can we expect extraterrestrial beings, if they exist, to see the world?

The search and the perhaps discovery of extraterrestrial intelligent beings are the most intriguing and revealing quest humans can pursue. Finding intelligent extraterrestrial life would confront us to a real "third party" vision of "out there." It would allow us an understanding of ourselves and the universe that alone we could never gain. Discussing extraterrestrial intelligent life is part of the Brain-centric program to make us as conscious as possible about our limitations and eventual possibilities to overcome them. It also helps us understand our place in the universe and thus gives a deeper meaning to our existence.

What we know about mental spaces already allows us to say quite a lot of things about these beings in case they exist.

But let's start with what has occupied us in this chapter: perception. Could any developed (intelligent) entities have "come up with" an alternative to getting knowledge about what is out there that is not using expectation mechanisms?

This question is crucial. Our senses work through expectations, what they render to our brain cannot be isomorphic to what is out there. But would isomorphism be possible for other developed species using different strategies than the ones that have developed on earth?

An "isomorphic" system would sense and picture information received from "out there." It would store these pictures in a database of analog pictures. If a picture is missing, no adaptation is then possible. The animal would not present sufficient complexity and adaptability for individual intelligence to develop.

We know that on earth mammals all use an anticipation perception system, allowing them to adapt and learn by trial and error. If they use an anticipatory system, they must have a natural world and thus feel.

Insects seem to use a direct representation system. With such a system, changing perceptual representation is difficult. An individual can get stuck in a "behavioral loop" and repeat the behavioral pattern. He can end up dying because of a minor change in his environment for which he has no corresponding existing pattern. Experiments with the Sphex wasp, for instance, illustrate these behavioral loops, as a representation does not adapt, the insect continues having the same deceitful behavior.

If insects survive, it is because they have another adaptation process that does not require learning. It is the Darwinian natural selection. At the species level adaptation derives from reproduction and natural selection.

Advanced animals adapt by natural selection at the species level and by learning at the individual level. This second level is because of the expectation mechanism changing the representations of their natural world.

Behavioral loops are not to be confused with compulsion loops, which concern advanced mental spaces. Compulsion loops are patterns, habitual chains of activities that repeat and provide a neurochemical reward such as dopamine release. The result, for instance, from practicing video games or using websites designed to gain customer loyalty through addiction.

Expectations and fill-ins exist in every advanced living species. It implies that these species have emotions and feelings and a natural world. No species could develop "individual intelligence" by only using direct isomorphism. Direct isomorphism has to be analog.

Humans, because of symbolic language, have developed a third level of adaptation by which we transform the environment to suit our needs. This third level happens only because we have sufficient complexity to digitalize and materialize our ideas.

Because of the universality of expectations, we can cheat a dog who could start running in the wrong direction when one simulates throwing the ball.

Let's consider a simple remark that highlights the difficulties humans have regularly come across in their quest for truth:

Only an external third-party observer can compare the representations of two individuals, but that third party cannot have direct access to A and B's experience and representation system.

Let us say A wants to compare himself to B. He will analyze B through the filter of A's representations. Same thing: if B is in charge of the comparison, he will analyze A through B's representation system. One will then get two results.

An external observer C having access to A and B's inner representations does not exist. Only A has access to A's representations, and only B has access to B's representations. So A, and B, cannot know how good their models are. Neither A nor B can be "correct." Humans have resorted to truth imposed by an authority or truth got through discussions and consensual agreement.

Since the 18th-century enlightenment, science has established a "third-party" description method of "reality" through narrative accepted and verified by experimentation. This third-party perspective is what we have called ERP.

Brain-centrism asserts that ERP cannot eliminate the intervention from the observer's perspective. We cannot create an external third-party C out of A or B by setting up rules. One can't eliminate the specificities of the brain's physiology that would allow us to build ERP isomorphic knowledge.

* *

Summary of chapter 1:

- Survival has been the key driver of our brain development
- The brain uses anticipatory systems specifically for perception
- Perception is subject to biases. Only personal biases can be corrected.
- When information is missing, the brain fill-ins
- Expectations and fill-ins are two driving motors of the mental space
- The mental space works by fragmenting reality and individuating objects out there. Language extends and reinforces fragmentation and individuation.
- We can divide the human mental space into two worlds: the original analog natural world and the more recent digital verbal world. We live in both.
- The mental space has two sources of information. The physical integrated analog Gödelian information from our body, the symbolic digital Shannon information from our sensory system.
- We associate every representation with feelings
- Representations form networks who are themselves representations associated with feelings
- The expectation mechanism is universal for intelligent species. Feelings and emotions are essential for intelligence.
- Mathematical objects are their verbal definition.
-

* * *

II: THE FABRIC OF THE MENTAL SPACE

> *The mental space is the inner place where we experience our lives, take our decisions, and build our tools to act on "out there." Two sources of information are available for the mental space to build representations.*
>
> *We have distinguished two worlds in the mental space according to what source of information is dominant. This chapter will further discuss the "building blocks" and the functioning of the mental space.*

2.1 Memory

The famous author Philip K. Dick is one of my favorite science fiction writers. He published in 1966 a short novel entitled: *"We Can Remember It for You Wholesale"* in which the hero becomes a client of Rekal Inc. He buys from Rekal Inc. beautiful vacations on the planet Mars. Not physical vacations, but "memory implants." Rekal Inc. guarantees that their memory implants are indistinguishable from life experience,[3] but at a much lower price.

Dick's idea has the merit to pose the question of "what is our identity," if not our memories. It also raises other questions: What is the nature of the past? What is a "true reality"? If one cannot distinguish our "real" memories from implanted memories, what is reality?

The sense of "being me," the feeling of continuity and permanence, merely manifests because of memory. We can feel and observe the "same body" and the "same feelings" today as last week, only because we can remember. If someone can implant or erase our memories, then who are we?

For now, Dick's idea is still science fiction, but will it remain so? For many neuroscientists, the answer is no; they are working

---

[3] Two films both named Total Recall have been inspired by Dick's short story. https://en.wikipedia.org/wiki/Total_Recall_(1990_film)

on ways to "read" the brain, and to implement or to erase memories.

Our memories define how we perceive ourselves and who we are. Based on them, we have built our "self-image," and our identity. Losing our remembrances, like with Alzheimer's condition, is losing ourselves; you feel yourself go. Later, the feeling itself goes. You are there, but gone! One can lose memories after a traumatic accident, a stroke, or an epileptic crisis. There are many types of losses. It shows we should distinguish various memories.

Memory losses are always scary. When you don't recognize anybody, you don't know whom to trust, you cannot even trust yourself. It is also traumatic for your close ones. Sometimes one can recognize a distant colleague, but not recognize one's wife or children.

Language remains memorized in most memory losses, alike when understanding of the words disappears. We remember music, words, and smells through different channels. Music, like language, is very enduring. Some Alzheimer's cases that have lost their souvenirs can still play piano or sing a melody. We remember smells when we come across them, but because they are not visual, we may have trouble recalling them. A general rule is that we remember details associated with powerful emotions.

When meeting a very different person, I often wonder what it would feel like living in his body with his mental space, his history, and experiences. I know how it feels to be me; (although I am surprised), what does it feel like to be him? Because of perception by expectation, two mental spaces that have observed the "same" scene have had unique experiences and will recall something different. The sense of self, the very core of all our knowledge, is something we can only know from the inside. We can only infer others have it, and we can only guess what it feels like to be someone else.

Other people's stories feed human memory today. Controlling the narratives has always been a major concern of people in power. Through narratives, one controls memories, expectations, and mental spaces.

As George Orwell put it, *controlling the past is controlling the future*.

The only way today to implement a narrative in a mental space is through language, communication, education and learning. Many[4] people estimate that the capacity of transmission is overly narrow. They wish to find some direct way to access our memory. Something like plugging a USB key into a computer. If such an invention becomes possible one day, we would, however, still not be at the level proposed by Rekal Inc. Such an implemented knowledge would only concern the verbal world, not the natural world governed by Gödelian information. We would know that this learning is not "real," it is not something we have lived! To involve the natural world in such a narrative would, however, perhaps be workable by adopting repetition and belief echo.

The metaphor "building blocks" we used to qualify as memories find its limits. Memories don't look at all like blocks. They look more like liquids that can divide into fresh streams or regroup in powerful rivers or hide in deep caves or even evaporate. What we have individuated as memory does not have precise contours.

One distinguishes two categories of long-term memories: explicit memory and implicit memory.

Explicit memory requires to be recalled consciously. We associate them with a time and a place; they belong to the verbal world.

Implicit memory is not about precise facts but is more of a semi-conscious, emotional recollection, it's a memory of the natural world.

---

[4] Elon Musk has also expressed this opinion

Children form explicit memories only when they reach the age of two, but most of their memories remain implicit until about seven years old. They can remember things before the age of three when they are small, but by the time they get older, they lose those early autobiographical memories. Sigmund Freud has called this phenomenon *childhood amnesia*.

One considers implicit memories as the pillars founding our mental space. They take part in the brain's formation already a few months after conception. Implicit memories are physical and analog and possess a large emotional content. We attach them to most of our representations and they allow us to remind ourselves. Because of their emotional origin, they can act on the body.

When we recall memories, the present state of the mental space influences the way we perceive them. Feelings, at that moment, do not always prioritize the same components of the memory. They put certain elements forward, neglect others, according to the general state of the mental space at the moment of the recall. The process is like what happens in perception.

For instance, at someone's funeral, one does not recall the same aspects of the buried person we would have remembered one year before.

Natural selection did not bother to create precise and stable memories. Most of our souvenirs are implicit, analog, and imprecise. Nature privileges speed and adaptability. We always associate feelings with representations to memorize them. When you pull your hand away from a hot cooking plate, it can happen instantly because of the analog memorized expectation. Sometimes the cooking plate is not even warm, but it triggers the reaction before you get to know it.

When we remember an ancient event, probably we remember a more recent recall of this event, colored by what these recalled have added or omitted. Over time, what remains of the event disappears. It is a bit like when you have played often an old vinyl recording; you hear more noise than music. It happens to any analog system.

Our brain also uses digital short-term memory. Let us say you have to remember a phone number. Without repeating it many times, humans can remember five to nine digits. For that, we resort to auditive memory. We use a different circuitry for visual short-term memory, which is even smaller.

For instance, when we read, we are shuffling things in and out of this digital memory. We build words not to exceed a certain length to facilitate their apprehension. The same thing when we listen to somebody speak. The limited sizes of short-term memory have enormous consequences not only on our communication bandwidth and capacities but also on our global understanding and our worldview.

We have difficulties processing an intricate calculation for which we have to memorize many intermediary results without a pen and paper. Not for a computer with plenty of RAM. When short-term memory becomes a longer-term memory, we associate it with an analog feeling.

Scientific American reports on the life of the savant Kim Peek, who is capable of unimaginable long-term memory capacities. "*His repertoire included the Bible, the complete works of Shakespeare, US area codes and zip codes, and 12,000 other books.*" He remembered every word of all these books.

In another article, Scientific American Mind discusses with Daniel Tammet. He is the author of two books, "*Born on a Blue Day*" and "*Embracing the Wide Sky.*" He holds the European record for reciting the first 22,514 decimals of the mathematical constant π (PI) in front of a jury and television cameras for five continuous hours, with no mistake.

These incredible performances of savants often come at a cost to them. However, they show we are still far from understanding how memory works and its capacities.

* *

2.2 Brains and mental spaces

Three hundred thousand years ago, the brain size in early Homo sapiens was already within the range of present-day humans. Studies show; however, the brain shape continued to evolve. It reached present-day shape between 100,000 and 35,000 years ago. We can surely consider that the brain of our savanna hunter was very similar to ours.

The relation between the brain and the mental space is a complicated problem. Some see the future in accessing the mental space by manipulating some brain hardware or wetware. As per 2021 we don't know, for instance, how the brain stores memories[5]. Certain neuroscientists dream about discovering a "brain code" that would allow a direct correlation between physical brain activity and mental space representations. By probing this activity and translating it through the brain code, one could then know what the individual is feeling, thinking, or dreaming. If such decoding of brain signals is possible for the motor cortex, as Miguel Nicolelis's BMI[6] shows, it is controversial for the so-called higher brain functions. A brain activity or neural patterns do not correspond to mental space activity.

Brain-centric considers that such a code cannot exist[7]. Feelings are central to any representation and are integral to any thought. According to brain-centrism, we express feelings in unreadable Gödelian information. Measuring brain activity is only about the

---

[5] The most widely held view is that memories are stored by enhancing synaptic connections between neurons. Recent experiments have suggested that memories may be stored within the nucleus of neurons, where RNA is synthesized.

[6] BMI stands for Brain Machine Interface. Is a technology to read and decode brain's electrical activity, with implanted electrodes or with a helmet, in order for a computer to execute the brain's instructions. Miguel Nicolelis has developed many application of the technology, for instance to help with spinal cord injury.

[7] Every few years, though, one can read in the news that a group of neuroscientists has read thoughts or dreams. To my knowledge, it is journalistic hype, or, in the best case, they are reading "silently spoken" words, the signals they consider result from the motor cortex activity.

digital Shannon information in the brain. That is the case for BMI's motor functions.

Brain-centrism refutes the possibility of a universal brain code. Such a code would imply an isomorphic correspondence between the brain's electrical signals and our experienced feelings, but also between brains and computers. The computer could read the brain signals, decode them, and print out your thoughts and feelings.

One could then describe the universe, including all mental activity, in the language of matter. Scientists that hope to discover such a code have adopted the "computationalism" paradigm. The brain would then be a mechanism, rather than an organism striving to remain far from thermodynamic equilibrium. (2.9) This paradigm is, according to the theoretical physicist Lee Smolin, responsible for us losing at least ten years in brain research. (5.13 and 5.14).

However, the brain engenders mental space. Our mental space must undergo the constraints enforced by the brains' physiology and structure. The environment has regulated the evolution of the brain's primordial parts associated with motor activity. More recent regions, such as parts of the frontal neocortex, have developed to serve the needs of the mental space. Once the verbal world became predominant, our brains adapted.

Among all advanced animals' brains, the human brain at birth is by far the smallest compared to its final size (less than 50%). We thus create considerable new brain matter, a phenomenon called neurogenesis. We correlate the brain's neurogenesis with mental space learning and memorizing. This underscores how crucial education will be for every life. Memory in the mental space thus corresponds to increased physical connectivity and neurogenesis processes in the brain.

The sequence of a child's brain's expansion is from the back to the front. The last region to mature is the prefrontal cortex that reaches complete expansion at 25. Before that, feelings and the natural world dominate the mental space, rational thinking is the last to develop.

Brains grow at the cell level by a choreographed process of cell proliferation, migration, and differentiation; but it also develops at the connectivity level with axonal growth, dendritic connectivity, and general refinement of the neural connections. For humans, this growth is a regulated, dynamic process that continues well into adolescence. For the mental space, it corresponds to the fastest growing and structuring phase.

The brain, and the mental space, are complex adaptive systems. Everything inside the brain is shifting, developing, retracting, adapting at every level from the molecules, to the cells, and entire brain regions. Actions and retroactions multiply between brain regions and between levels. This happens through electromagnetic, chemical, and physical forces. It provides the brain its marvelous adaptation capacity, named brain plasticity, and is critical to learning and adaptation. Plasticity is ongoing throughout the lifetime and so should learning and memorizing.

The connectivity of the cortex corresponds to patterns of neural activity. We create these patterns through novel representations of the surroundings. The brain's growth and adaptation correlate to the quantity and the diversity of new representations generated in the mental space.

Scientists have shown that physical exercise helps neurogenesis and plasticity in the hippocampal[8] region. Sleep plays an important role in dendritic growth. It reinforces connections and favors long-term memory. Many studies show we can improve long-term memory of adults by training.[9]

---

[8] Hippocampus is a brain structure embedded deep into temporal lobe. It has a major role in learning and memory.

[9] The method of loci is a common technique used by memory sports competitors. It refers to the memory enhancement strategy of associating the items to be memorized with specific physical locations and retrieving the items by mentally "walking" through the imagined place. It is an old technic, its origins are traceable to ancient Greece and Rome

However, certain areas of the brain are not as plastic. Even regions that play critical roles in factors such as movement, language, or speech. Damage to those areas can cause permanent handicaps, as other areas cannot take over.

* *

2.3 Computer's mental space

Even if some parallels exist between the computer's memory and the brain's memory, they function differently.

We did not design computer digital memory for the same usage as human analog memory. We require it to be accurate and to repeat the result each time we need it. Computers would be useless if their memory were like ours.

It is confusing that computer scientists have the habit of employing names attributed to brain functions. They describe computer attributes which "seem" like brain functions but are de facto different. It is the case with memory, but also intelligence, decision, and many other words.

A digital computer has a mental space when we switch it on. At the difference with an organism, one can switch it off, cut the energy source and when we switch it back on, continue working as before. If an organism is "switched off" it dies and the process is irreversible. Time reversibility is a critical difference between a mechanism and an organism.

Laws of physics are (locally) time-symmetric; one can change the time variable t into -t in the equations and nothing special happens. Because the effects of mechanical and electromagnetic phenomena, as described by the equations, are time-reversible, one can "wind back the clock" and restart any experiment. Without this local reversibility[10], the universe would not be understandable and we would not exist.

---

[10] Time reversibility is a mental abstraction. Laws of physics consider isolated objects that allow this reversibility. They don't apply to organisms who are dissipative entities. (2.9)

We cannot consider a computer as a third party as we would consider an intelligent extraterrestrial. This is because it did not evolve, we built it as a (very limited) projection of our mental space. We projected on him our individuations, our fragmentations, our mathematics, and a large part of our verbal world. However, as our natural world works with integrated, subtract dependent Gödelian information, we could not project it. He is a "one world" mental space lacking the complexities of the two worlds in which we struggle. It then means that he lacks intelligence and does not understand. When computer scientists use the same word for computer and human attributes, they "neglect" the natural world representations and feelings attached to it. Those attachments give life, meaning, and purpose to human words. By attributing these words to computers, they strip them of their essential substance[11].

Examining some differences between his memory and our will give us a glimpse into the computer's limited mental space.

1- The brain's memory is contextual; how and what we recall depends on the present state of the brain. The current status gives its "color" to what we remember. It uses a fill-in process, adapting the memory to the present situation. Computer memory is a precise digital copy, independent of any context.

2- Brains' memory does not have an "address." It instead works by associations and analogy; it is implicit. When one tries to remember something, one evokes the history or the context in which it happened. Many routes, many associations allow us to recall a name or an event. We do not search in our memory as we explore in a directory by knowing the address.

3- The brain's memory is not precise; it is not dealing with numbers or data, but with associated feelings and impressions. We remember what we have experienced rather than measures and data. We remember something like "She was a pretty big

---

[11] This sneaky mechanization is the main subject of my book *L'ordinateur ne digèrera pas le cerveau,* 2013 Editions Sarina.

and imposing woman," rather than "She measured 1,925 me and weighed 124.56 kg."

4- When we recall a memory, we color it with the present context as per point 1-, when we memorize it again, the recent recall alters the memory. We recall having remembered. Memories of old events can this way get very distorted by successive recalls and fill-ins. Computer memory is an exact digital copy.

5- Brains forget. One can erase computer memory; they don't forget. Brains "believe" something happened; they can think, although they are not sure. Computers have something in memory or do not have it. They cannot believe or doubt. They have no underlying feeling.

6- Brains can have false recollections, illusions, a computer cannot. Computers' "false memories" are correct when compared to the input. If its memory contradicts with facts, the input already contradicted those facts.

7- Human memories influence perceptions by the expectation process. The computer's memory content does not affect the inputs.

8- Not only we remember sensory data or verbal information; we remember in priority smells, feelings, imagined stories, dreams, fears, beliefs, trust. We memorize implicitly and explicitly. Computers do not have feelings, they do not trust or "believe."

9- Memories change the "hardware" of our brain. Plasticity and neurogenesis create new molecules, synapses, and pathways. The volume of the hypothalamus, or the white matter, can vary. If we stop using a pathway, the memory vanishes. It is not the case for computers.

10- Two human witnesses will never describe the same event in the same way. They have not perceived it the same way; they did not have the same expectations while observing. The state of their brains was different. Two computers with the same inputs will produce the same output. If they do not, it means that we have programmed them not to do so.

11- We can encode a computer's digital memory on different IT devices; it is not the case for human memories. We already noticed that our analog memory is physical and causal; we cannot detach, and is not directly communicable. We sometimes "know," but we cannot explain it with words, the translation into digital communicable form is difficult. We often need an analog picture or a drawing to communicate. Human language is imprecise to describe our feelings, for instance. What we memorize by "doing" cannot fit into words. One cannot explain a color or a taste to someone who has not experienced it.

12- Remembering a piece of music can make you cry. Remembering a situation can make you shiver. This is because memory information is analog and integrated. Computers do not tremble when we load a file. The mental space of a computer has no natural world. It has no imagination, no understanding, no creativity, no identity.

The precision and rigidity of the digital computer, the reversibility of its operations, make it useful. It does not make it intelligent.

**

2.4 Mental Patterns

The brain is not only constantly expecting; it is also constantly building or adjusting models. Model building and adjusting are like a motor permanently switched on, we can hardly stop it running and producing combinations, plans, micro-realities. Our models are combinations or successions of mental representations. It is like a game that never stops, we cannot put it at rest. The mental space produces models not only because they are useful to solve a problem, but for no reason, only to play. Models use whatever representations are at their disposal at that moment, colors, shapes, equations, musical notes, sounds, smells, concepts, any building material will be fine as long as the game goes on. Often we call thinking this act of model building. Because of this we build stories, we imagine situations, we prepare for the future. Even when no sensory input feeds new representations, the mental space will continue this vital activity. During dreams, deprived of any censorship and sensory inputs, these constructions go on. During the dream period, the neural activity in the primary sensory areas of the neocortex produces the impression of sensory perception.

The capacity to coordinate this choreography and make it purposeful is critical, at least at certain moments. With a controlled choreography, a natural game of our mental space can become some useful creation, some answer to real questions, or some plan for the future. If the motor seems to be always on and the vehicle always running, most of the time it's going nowhere by lack of guidance.

Many of these building patterns end up being repetitive. Either we build the same thing or we use the same pattern with other Lego pieces to build something else with the same model, in an unconscious and perpetual mental ballet.

Mental models will also induce behaviors. The mental activity is similar for these behavioral models, they repeat during our entire life, and often without us consciously noticing it. Certain thinking processes, certain feelings are like routes that one repetitively follows, like unchangeable habits.

When these models are so deeply rooted in implicit memories and expectations; they serve as a "mental background" ready at any moment to take control of our mental state in response to external circumstances.

These mental background patterns avoid the difficulty of thinking, they are ready to use on the shelf. However, they influence expectations and determine the way we "automatically" perceive a situation and respond to it.

For instance, one will not think, react, or behave the same way at the office, at a wedding party, or on the beach. Each situation has its unconscious and usual behavioral patterns. Even some minute events can unconsciously trigger patterns: a facial expression, a smell, or some analogy with past circumstances can have us recall old and forgotten emotional attitudes. Behavioral patterns triggered by external circumstances often drive behaviors like being shy, not knowing what to say, being uselessly aggressive, for instance.

Stereotypes are typical thinking patterns; they are ready-made, ready-to-use thoughts or reactions, avoiding the thinking effort. They have the advantage of making the person more predictable. For instance, with a stranger, we start by carefully speak about uncontroversial common ground: the weather or the beauty of the landscape.

By "feeling" what pattern has adopted our interlocutor, we adjust our expectations accordingly and expect his answers. Social and work conventions establish communication patterns and codes that rule professional relationships. Our behavior feels then like pre-programmed. Everybody knows what to expect.

Many patterns are conscious. In a classroom, at the border, at a concert, at the restaurant, we adopt conventional stereotyped behaviors; they facilitate social relations. Everyone "plays" his reassuring conventional role, the same way children play being a doctor, or a police officer, or a superhero.

We learn and memorized deep patterns through explicit and implicit memorization during childhood. Although very influential, they are difficult to change because they mostly remain unconscious. We may unconsciously repeat these types of patterns, even if they are harmful to us or others. For instance, the reaction of blaming others for what happens to us is frequent in childhood, but we should not let it turn into a repeated pattern. We also build other deep patterns upon ancestral fears. Racism falls in this category.

Repetitive *behavioral loops* happen when the individual has no way to observe, interpret or understand the result of an action. If this process produces any reward, he will go on repeating it frequently with tragic effects. We have investigated behavioral loops in insects such as the Sphex wasp, whose perception involves no expectation mechanism. No adaptation can "correct" or interrupt the loop that continues to repeat like a bugged program.

Patterns are specific to each mental space and its history, and they interfere in every aspect of our life. Each of us, for instance, makes his resolve to questions like.

When is something too risky? How much can I disturb someone else? What does it mean to be cautious? When do I feel ashamed? How much should I fight back? When do I decide that this be too much for me? How do I show I have reached some limit? What is unbearable? How much am I expecting from myself? How much do I blame others? When should I feel a situation is unfair? When can I say, I have understood? How can I pretend to be what I am not? How much can I cheat, lie, or betray? When can I feel, it's none of my business? What makes me feel responsible? Do I have duties towards society? Do I feel I have a guardian angel? Do I over-trust my feelings? What is my interest?

The diversity of our answers is a richness for human society, something that seems impossible for one of us will be an attainable goal for another.

Some of us end up behaving in such rigid conformity with social patterns they forget what it means to react spontaneously. They have buried any natural reactions that their natural world would suggest. These behaviors end up being very damaging for the mental space and social interactions. One is at risk of losing the sense of purpose by not allowing oneself to think and feel, to develop one's own opinions and beliefs, and express them.

Behaving, as an adult, does not mean we have to forget how to be spontaneous and creative as a child.

* *

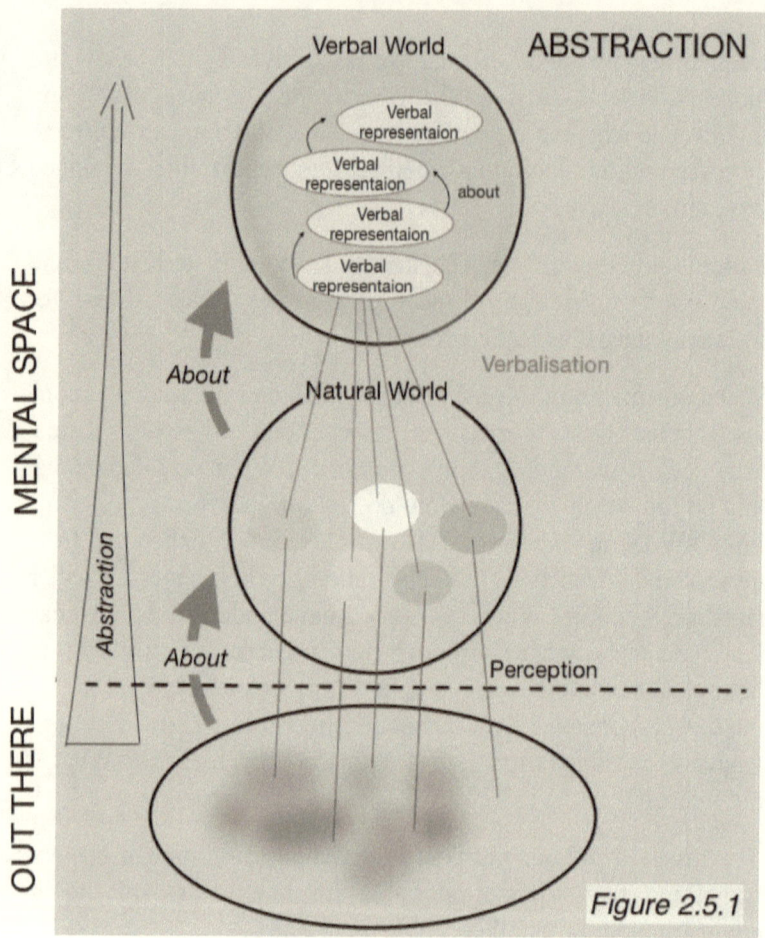

Figure 2.5.1

2.5 Mental abstractions

We shall pursue now our exploration of components and functions of the mental space with one of its most important capacities.

The term "mental abstraction" is tautological. Abstractions are mental! Nothing is abstract "out there," abstracting is an activity of a mental space seeking to organize itself. Abstracting corresponds to structures of the neocortex[12].

---

[12] The following sentences are obvious, however, we may have some difficulties fully understanding them as we are so deeply engaged in the ERP paradigme.

Abstracting creates a new representation, a new concept, new individuation "about" a group of representations by adopting "analogy." The analogy can involve a subpart or a shared property of all the representations of the group.

"Vertebrates" individuates a subgroup of animals who have an "analog" spinal cord. "Vertebrates" is a mental abstraction that merely happens in the mental space, not out there.

"Family" individuates a subgroup of humans who have a close genetic relation (the analogy). "Family" is a mental abstraction that only happens in the mental space, not out there.

"Face" is an abstraction derived from similarities between some parts of Paul, Mary, a lion, and a dog, a painting... We have individuated something analogous and abstracted it with a new representation that we then called a face. A face is something "about" Paul, Mary, a lion, a dog, a painting... A face cannot be something out there. Abstractions are always mental.

The new abstracted representation will thus be at a "higher level of abstraction," "about" what it abstracts and is therefore sometimes called a meta-representation. Abstraction thus creates hierarchies of levels, where each level is about the underlying one. The critical word here is "about." Being "about," means it does not belong to the same category of representations. Gravity is about massive objects; it is not itself a massive object. A crowd is about people but is not a person. The set of all sets is about sets, it is not itself a set. The hate of hate is about hate is not "hate." A forest is about trees, it is not a tree. The verbal world is about the natural world, it is not the natural world. A physical law is about what we describe, it is not what it describes.

"Out there," the "natural world" and the "verbal world" represent therefore three abstraction levels, each being about the one underneath. (Figure 2.5.1.) "Out there" is the ground level, it is not mental[13]. Our physical emotions also represent a ground level.

The ground level is not "about" anything else, it is itself. We can't know much about its nature, we can only know how we represent it in the natural and in the verbal worlds. Kant calls it the "noumena." Einstein and Hawking called it the "mind of God."

The natural world is about these ground levels. It collects information through perception to build its analog representation network. We described how these perceptions use an expectation mechanism. The organizing principle of this world is the analogy.

The verbal world collects information from the natural world and conveys it through digitalized symbolic language. It cannot be "identical" or isomorphic to the natural world about which it is. The structure of language allows us to continue building abstraction layers within the verbal world. A word can be about a word and a sentence can be about a sentence. The verbal world can therefore build abstraction towers. Each level is about the level underneath and cannot belong to the same category as it. The organizing principle is causality.

Most of the time, we are not conscious that we are abstracting; we are so used to live in our abstractions that we forget these operations are taking place in our verbal world. This powerful property of the mental space allows us to create unlimited hierarchies of abstraction, boosting the complexity of the mental space.

But abstraction has another important characteristic: when we have selected the similar part in the representations we abstract, we have also neglected all the parts that are not similar. Out there, such fragmentation in parts cannot exist as we examined in the first chapter.

---

[13] Rather, it's the mental representation of "something" we believe not to be mental.

This mental operation of ignoring difference to create higher-level representations turned out to be critical to building our realities. It granted us to apply thinking processes to large categories of represented objects. It allowed us a shallow understanding of large categories. Without neglecting some characteristics, an object would not even be identical to itself because of the internal ballet of its molecules. We would have an incredible number of individuated categories to which no general rule would apply.

For instance, one can assert that *"dogs are friendly animals."* We don't have to go animal per animal or dog by dog to assert their friendliness. By neglecting differences, we lack precision but gain understanding at the cost of finding some exceptions. Savants who individuate and memorize every detail improve difficulties in abstracting and painting a larger picture.

Money is an abstraction that covers many "things." One associates a dollar value to nearly everything. However, it says very little about the objects from which it is abstracted. As we express it with a number, it has a lot of weight in the mental realities. A balance sheet gives a picture of a company at a moment, but it's like a picture taken with a monochrome camera. It neglects a large and important part of the information that could change our appreciation.

Representations in the natural world are first-person perspectives, we cannot measure them from the outside and remain personal knowledge. Representations in the verbal world use a common language and favor exchanges between individuals that know the language. However, only the words and what they represent will be common, not the feelings associated with the words. Different people will then have different interpretations.

* *

2.6 Orders and invisible worlds

> *All effort to bring order in disorder is disorder.*
>
> David Bohm

Abstracting never stops. As expecting and filling in our mental space cannot avoid it, we also consider abstracting as a driving motor of the mental space. One of the key abstractions is generating models or "orders."

We enter an office and we can instantly feel how the owner has established it. When we observe a person carrying out a task, we can figure out how well this individual organizes himself. The organization of a city, an office, a person, a succession of movements is a concept we abstract from the relative position of components of the scene. The organization is a synonym for the optimization of certain values. We organize to save time, to remember, or to save space or energy. Any organization follows logic, a system, an algorithm. It can be a natural logic that everyone understands, like arranging books by authors and authors by alphabetic order. It may be another logic that the owner assumes.

Our mental space abstracts all the time. Higher-level abstractions become part of what we expect to see. Over time, we memorize many models of organizations. Each appears as a higher-level representation we call an "order." Orders are not only abstractions of the verbal world, but we associate them with feelings in the natural world. Discovering an order, where before we saw nothing, is for most people a satisfaction. Discovering an order brings us some clarity and understanding. Certain orders are obvious like axial symmetries or by size classification. Others require much more study.

Science aims to discover orders. A scientific law can exist only where we perceive a sufficient regularity, that means an order. The law itself describes how this regularity works. Science does not apply in regions with any regularities.

Many mammals can memorize orders, they seem to understand an event follows another and can abstract an order. Such a succession of events becomes a model that we will use as an expectation.

Not every one of us individuates the same orders. What appears as an order for someone will go unnoticed for someone else. A good automobile or plane pilot expects noises from his machine. He has trained to become sensitive and correlate multiple events and knows what sound a correlation should produce. A novice passenger would notice nothing.

Orders can become expectations, they are part of the perception mechanism. Richer, the library of orders an individual has made up, larger will be his explanatory reservoir.

Our search for orders is thus a search for understanding. It fills up our need to find simple rules explaining complex natural phenomena. This simple rule is often very hard to find. Orders are not immediately visible. But these invisible orders are the only way for us to understand the visible.

Discovering orders was traditionally the entrance door to a secret, invisible world.

In chapter 1, we have described the two sources that feed the mental space with information. The body and its emotions and the sensory information from out there. Many humans feel that a third source also feeds them. This third source would provide us with intuition, guidance, deep convictions, and meaning. The information would also be Gödelian. Once translated into the verbal world, people call it God or religiosity, Nature, or the Universe. We can only know this Gödelian information from the inside. Any third-party description or explanation in the verbal world will not be faithful. Brain-centrism accepts the idea of this third source. However, we cannot distinguish it, as a third party, from the emotional source, as both are Gödelian.

**

2.7 Cave painting

Let's imagine we are now in the upper Paleolithic era, some 25,000 years ago somewhere in Europe. Homo sapiens has remained as the sole human species alive on the planet after the last Neanderthals had disappeared 10,000 years before. A man from a nearby hunter-gatherer tribe, a gifted Shaman who has learned how to master different states of consciousness, is wandering in front of a large hole in the rocky mountain. He has taught himself how to master his emotions and develop empathy to understand the feelings of his fellow humans without letting their emotions invade him. Those skills have earned him the position of Shaman.

The night before, he had a series of dreams; he has meditated and felt that these dreams were essential messages. Dreams have enormous importance for him; without a structured symbolic language, he only has his emotions to interpret them and limited ways to communicate them to his fellow humans. Cave painting was his way to express these visions to every member of his tribe.

The mastery of his fears and his desires, he has developed, freed up mental energy, allowing him to dedicate forces to higher mental activities. He learned how to control his expectations and his moods. The wisdom he gained gave him many social responsibilities in his tribe. His knowledge of plants has put him in charge of the health and the souls of his community. He is free from the obligation to go hunting, and his tribe values his capacities they take care of his basic needs. What he is about to do today is part of his social duty. The shaman is the only one in his group who has time to "think" and to meditate. After analyzing his dreams, he attributes them to an external force, a divine entity expecting something from him. (The third source of information.)

He has studied for years the invisible world where knowledge hides. Initially, the old shaman guided and instructed him. He knows the divine creatures who populate that world, and the way they send messages he can interpret. He knows the healing power of the plants he collects for his sick fellows, and how to use them by respecting the hidden forces.

Figure 2.7.1: Reproduction of Pictures From the Lascaux Cave, by Maila Rinaldi

For him, the invisible world has become more important than the visible one. He believes his intimate relationship with the gods gives him his powers. He can master and interpret their symbols better than anybody else in his tribe. That is why he is also in charge of the burials when someone in his group dies.

He is now ready to translate his dreams into an artwork he is about to paint. He has found an obscure, hidden, and buried place, a rocky cave where he will work.

All over the world, wherever Homo sapiens lived, some 35,000 years ago, one can observe an explosion of cave paintings. From Australia to Asia and from the Americas to Europe, in a short period, humans started painting deep caves. It is impossible to say what made them undertake these dark and dangerous journeys. What thoughts and beliefs motivated them to place their images in the innermost, darkest recesses of the caves, only illuminated by the flickering light of an uncertain fire. However, this period, about 35,000 years ago, is like when the size and the shape of Homo sapiens' brain reached the ones of present-day humans. (2.2.)

It is not astonishing, considering what we know on an analogy, that the subjects of these cave paintings are always animals typical of the region, and more rarely, humans or vegetation. Religious motivations behind the cave's artworks were first suggested in the 1970s. The hypothesis was very controversial. It has now become widely accepted. For the first time, in the upper Paleolithic age, "analog material projections," of the human mental space, were taking place on the rocks of deep caves. The mental space has reached a sufficiently developed stage to project itself outwards in artistic works.

These caves illustrate the ancestral origins of human questioning. The same questioning is still ongoing today. Questioning is always about the unknown, the hidden world, the unreachable, things that our senses do not immediately communicate to us. The mystery is the root of questioning. As mysteries are invisible, we represented them with visible symbols. Little has changed. Science today still needs the invisible to understand the visible.

Other human species such as Homo erectus, Homo neanderthalensis, or Homo habilis, did not paint. Why did art expression suddenly start nearly simultaneously all over the planet?

"The remarkable thing is not that the shamans of the Paleolithic had the visions they did but that they accepted them as divinely or externally inspired, and that accepting them they translated them into mortal form, the art of their rock paintings." Says Dr. Ilse Vickers, who has for many years, pursued a professional career with the European Commission and with University College in London. She continues by writing:

"When we try and dissect the Paleolithic cave paintings and state that they either mean this or that, we have unraveled the precious and complicated weave of strands, and their true

meaning is lost: their treasure lies in their totality.[14] *The symbolic language of the unconscious is at the same time a most primitive language, arising from the remote, bestial instinctive spheres of our being, and it is also the language which gives man insights of the highest order, far beyond what our rational mind could produce by itself."*

The mystery is the source of arts. Attempting to make the invisible visible to everyone is what artists do, allowing us to connect and inspire meaning.

Carl Gustav Jung, the famous Swiss psychologist, writes:

"Such being the case, we should not be in the least surprised if the empirical manifestations of unconscious contents bear all the marks of something illimitable, something not determined by space and time, in short, something that has the quality of the numinous or divine."

The common brain physiology induces archetypical similarities in our mental spaces. That could explain why we find similar mythological motifs and symbols in religions of all ages and at places far apart.

In the eyes of modern science, certain beliefs can sound nonsensical. However, they have their role in other significant aspects of our mental space.

Shamans inspired respect because they opened doors to the invisible and answered open questions.

Studying humans' mental space will imply understanding how its most primordial projections led to its highest functions. It is not surprising that art and religion are so related; they both refer to the mysteries of hidden but deeply experienced worlds. They refer to those regions of the mental space fed only by inner Gödelian information.

---

[14] We expressed this idea by the difficulty to translate Gödelian felt information into Shannon syntactic information.

For our shaman and many of his followers in human history, orders were an entry door into the hidden world of the gods, a world that explained why the visible one behaves as it does. By asking questions about the visible world, anyone could feel the invisible world: Who created what we see? Why is it raining? Who made the world so beautiful? Why is it so unfair?

With mathematical objects, numbers, and geometry with their infinite variety of orders, mathematics became a mother tongue, the language of the gods. A necessary entry door to the invisible kingdom.

How could anyone doubt that behind the visible, there is the invisible? When seeing the sunrise every morning, by observing the cycles of birth and death, no doubt could remain; a hidden world had to exist. When discovering that they are only five regular polyhedra, or that the length of the hypotenuse of a strait triangle is irrational and non-countable, could one doubt mathematics was part of the mystery? Could one question these rules were the hidden manifestations of the gods? Our highest mission was then to try to "read their minds" and accomplish their will.

Cave paintings were just the first steps. A succession of incredible monuments has followed. Cave paintings, pyramids, temples, cathedrals, moon rockets have all the same origin in our mental space; they all materialize our quest for the unknown. They express our mental orders, our hopes, fears, and beliefs, projecting aspects of the mental space into the matter. Not all materializations follow the same pathway from an idea to its encoding into matter. Arts follow an analog path, technological objects follow rather a digital path. The frontier is sometimes, however, not as clear as I expressed it.

**

2.8 Cybernetics, adaptation

For hundreds of thousands of years, our organism has adapted to changing surroundings. Environments have molded us through heat and cold, through abundance and scarcity. We have survived, although we have gone through rough periods. The Toba catastrophe has reduced the total human population to some 10,000–20,000 individuals. That was 75,000 years ago. Brains who developed to coordinate and regulate movement have played a decisive role in our constant evolution through natural selection. Natural selection needs a "selection pressure." Currently, this pressure does barely come from an independent nature. It also develops from our constructions. We are closing a risky loop. We are adapting to our creations.

In the last 15,000 years, the adaptation process has reversed. Instead of only adapting to the environment, we started using our brain to transform it to fit our needs. We have developed technologies and cultures which shape the world to reduce natural selective pressure. We are the only animal that has transformed the environment on such a large scale. Our new environment has become closer to a projected image of our digital mental constructions.

A regulation requires two independent parties, one regulates the other adapts. What happens when the regulator becomes a creation of the party who adapts?

There is a difference between an analog (natural, biological) adaptation process and a digital one that requires a mental space.

The mathematician Norbert Wiener (1894–1964) presented in 1948 a powerful concept that he called cybernetics. He described it as "the scientific study of control and communication in the animal and the machine." Wiener coined the term based on the Greek "kubernētēs." He was not only a mathematician but also a philosopher whose book: "The Human Use of Human Beings. Cybernetics and the society" influenced generations of readers.

Cybernetics formalizes the notion of "feedback" as the mechanism of regulation and adaptation of a system. The first theorem of Wiener's book is a very general assertion of Ross Ashby and Roger Conant. It states:

"Every good regulator of a system must be a model of that system."

The tempo mat of your car is a regulator. It controls the speed by measuring it and acting on the gas pedal to adapt it to the desired value. The hypothalamus regulates our body temperature.

The theorem asserts that any maximally successful regulator must be isomorphic with the system it regulates. One also calls it *"the internal model principle"* or *"the good regulator (model) theorem."* One can picture it with the following expression:

"Every good key must be a model of the lock it opens."

The Ashby-Conant theorem explains why a brain develops a mental space to operate. To regulate the body and adapt to the environment, it needs to build models or representations. It also asserts that a computer has a mental space.

The second theorem, known as *Ashby's law*, is so fundamental that it's also called "the first principle of cybernetics," it states that

"the degree of control of a system is proportional to the available information."

The more one knows about a system, the better one can manage it. The more an entity collects information from its environment, the more it can control it or react to it. The first principle of cybernetics explains that a species that wants to maximize its survival should gather as much as possible knowledge about its environment. It supports the concepts of research and education.

Let us now examine the difference between an analog and a digital regulator. The good model theorem is about the "collaboration" of two entities: a regulator R and an object O.

With a digital regulator like a tempo mat, R collects digital information on the state of O, processes it, and exercises a physical action on O to guide O towards the desired state. This involves a reader, a processor, and a writer. The processor must know (or decide) what should be the optimal state of O he wants

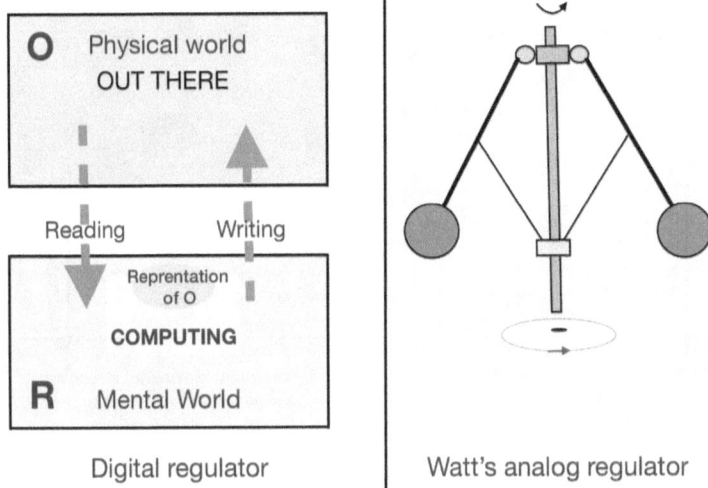

Figure 2.8 1: Digital and analog regulators

to reach so he can steer O in the correct direction: the *goal G*. A digital regulator thus needs a mental space, the regulation is then a process in which information and energy navigate in a feedback loop between physical and mental. It is a programmable device, one can change the value to be stabilized.

An analog regulator does not need a mental space and has no *goal G*. It uses the physical forces. The James Watt regulator (figure 2.8.1) is a classical example of an analog speed regulator that was used on steam engines. A planet orbiting the sun self regulates, no information to collect, no prefixed angular momentum, nothing to compute. Only Gödelian information manifests. *Planets do not know Newton's laws*. However, they orbit the sun.

Analog regulators appear everywhere in nature, in physical and biological systems. We refer to these regulation processes in physics with the idea of "conservation laws," such as conservation of energy or electric charge. In biology, we refer to them as homeostasis principles. Nature "out there" does not produce digital regulators, only mental spaces do, it has no pre-defined goals.

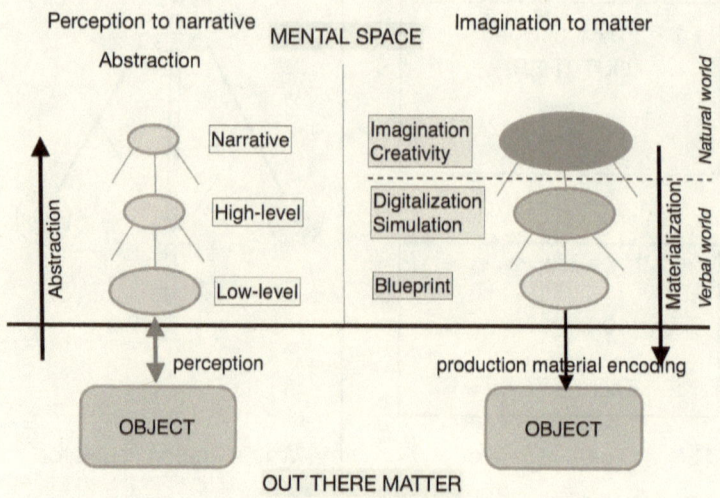

Figure 2. 8 2. Abstraction and Materialization

To our best knowledge today, we are the only entity in nature capable of producing digital regulators. How is it we can produce arrangements of atoms, objects that would not have evolved by themselves under the forces of nature?

Figure (2.8.2) represents the abstraction and the materialization process with two arrows in the opposite direction. For the materialization arrow, things start in the natural world with abstract imagination and creative images. Once these ideas have matured, they get digitalized, decomposed in parts, put into mathematical formulas and data. We are now in the verbal world, and we can run simulations. Once satisfied, we produce a blueprint that is a model of the ultimate object according to Ashby's theorem. What started as an abstract natural world creation finds itself encoded in the blueprint. From there we can transfer it, encode it, or print it into the matter because Shannon's information is independent of the subtract.

Natural	Human
Existing matter	Existing matter
No conscious representation	Conscious problem
Analog regulators	Analog or Digital
No model	Analog and digital model
No blueprint	Analog-digital blueprint
No Goal G	Goal G
Combined action of physical forces, evolution	Encoding into matter following the blueprint
Recombined atoms	Functional product
No reverse engineering	Reverse engineering
Analog causal effective information	Digital, non-causal effective information

Figure 2.8.3: Natural and Human Productions

Technology would not be possible without this digitalization process. It is what we do when we build a car, a computer chip, or a building. We start by preparing a thorough digital representation of our mental abstraction, a series of instructions describing step-by-step how to manipulate, transform, and assemble various pieces of the matter: a "blueprint." This blueprint serves as the printing program to build our object by "encoding" the information into the matter.

We can distinguish a natural object from human production because our productions go through a digitalization phase and natural objects don't. We fragment and decompose in independent pieces that we then assemble. That's why one can *"reverse engineer them."* We cannot reverse engineer biological entities, they have never been engineered. They have not been built by an external third party, they have evolved from within.

Nature does not build mechanisms[15]. Mechanisms and organisms are thus very different entities.

What we built results from our fragmented and decomposed thinking process. The verbal world cannot think otherwise, it has to use Shannon information. Out there is not fragmented, fragmentation is only the way our mental space works.

With technologies and digitalization taking over, we are transforming our environment. We are "materializing" our visions. Rearranging the atoms around us as a projection of our verbal mental world. Many of us only see, hear, smell, touch human productions all day long. At each step we are silencing our "natural world[16]" and forgetting that our vital sources are in there. It is different to do something because we have to follow a procedure or because we want to.

Confronted to new technologies, our older answers to ontological questions become irrelevant. We will have to rethink them. Technology itself won't solve our problems. Our human solutions are in the natural world, not in the verbal world that generated the confusion.

Our natural world is suffering. By adapting to our digital products, we are at risk of losing focus on what makes us human and engaging in a funnel leading to mechanization. We are losing our inner guidance. We behave as if the perfect human should be a robot.

The race to rationalization and productivity is enslaving us more than it is liberating us. Our evaluation instruments omit or cannot measure what is important, because it is not Shannon informational. The deeper questions are not technological, but about applications of technology. We should stop asking, what

---

[15] We sometimes call mechanism a production of nature, but in fact it has not been built as a mechanism it has evolved.

[16] In my book L'ordinateur ne digerera pas le cerveau, I warn against this growing mechanization of humans and the society.

can we produce, how fast can I produce? And quest for what is desirable to produce?

Knowledge and wisdom must return at the steering wheel, and technology must remain a tool, amazing, but a tool.

Digital regulators have given us the fantastic power to control limited aspects of nature. We regulate immediate and local aspects to discover later we're disturbing other, more important, equilibria. We navigate, guided by the smell of things that we created as tools but are becoming our masters.

* *

2.9 Adaptation, Complexity and Russian dolls

Let's imagine a series of Russian dolls. The dolls represent complex structures of increasing sizes. From molecules to animals, the planet, the solar system, the galaxy. The complete universe is always the last doll of any Russian chain. The

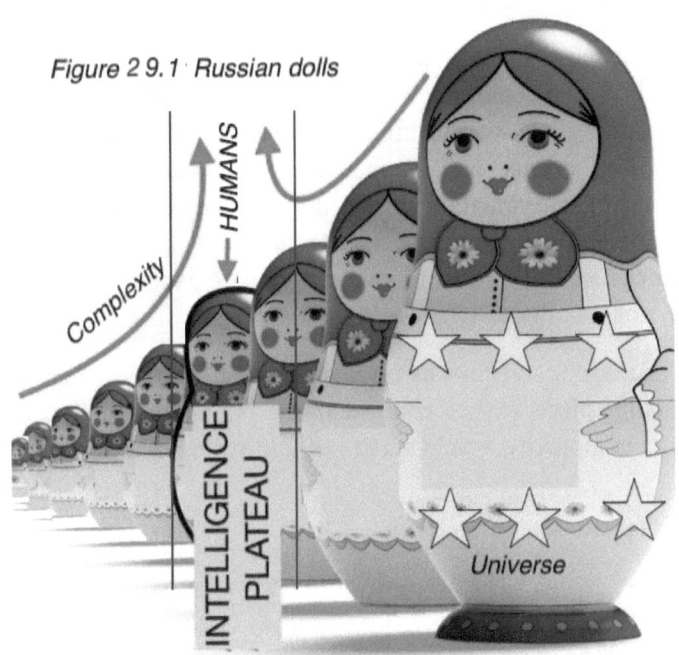

Figure 2.9.1 Russian dolls

smallest dolls are, according to 2021 physics, electrons and quarks.

This metaphor is a typical mental fragmentation. "Out there," does not separate the dolls.

The Universe is expanding, the structures inside it must adapt. Multiple waves of energies flow down and back up the Russian doll's chains, transforming the aspect and the structure of each doll. The Universe's expansion reflects on the galaxies, who, in their turn, affect the solar systems, who influence the planets. Actions and adaptations happen down and back up the chain at every stage, using any of the recognized physical forces and at specific time scales. A continuous feed-forward and feedback of activity spread through the full system.

Our Russian doll is a metaphor for the abstract concept of Complex Adaptive System[17]. That is a system with many levels of independent agents acting on one another. It could be a galaxy, a brain, the economy, the markets, … The mental space abstracts such structures to organize and make sense of its fragmented representation procedure.

When we individuate a doll at whatever level, we separate it from the entire chain. We can therefore not get a full understanding of its dynamic because of Ashby's theorem. Unexpected events will happen that originated outside the perimeter of individuation, in the dolls above or beneath. Lack of information explains the events that we call random like coincidences. Such randomness can only exist in the mental space, not "out there" where nothing happens.

As humans, our size puts us in the middle of the series. This is a fascinating place to be. Had we been bigger or smaller, our average complexity would have been lower[18] (Figure 2.9.1.). We

---

[17] We will further develop this concept in (3.8).

[18] We are comparing here a subsystem to its external environment (excluding the subsystem itself) like in unpacked Russian dolls (Figure 2.9.1.) If we do not exclude the subsystem, the complexity in the chain would be stable or increases from the smallest entity to the biggest.

are at the privileged place where complexity is sufficient for mental spaces to develop and invent structured symbolic languages. We will call it the *intelligence plateau*. (Figure 2.9.2.) From that privileged place, we have developed our third-party visions and communication. We have built digital regulators that otherwise would have never existed and materialized ideas into structures that the universe does not produce directly himself.

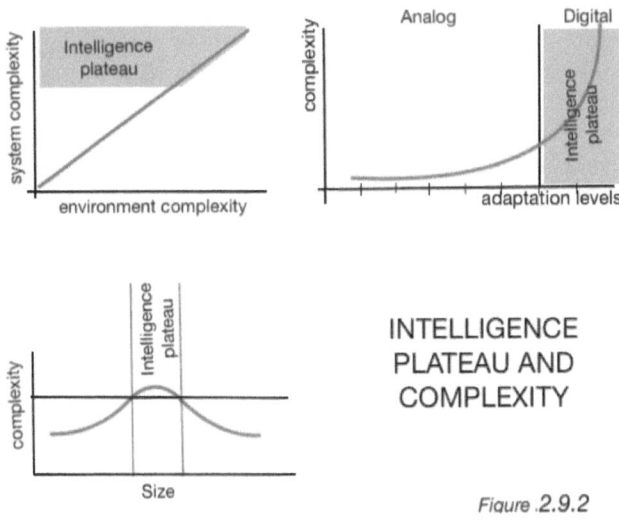

INTELLIGENCE PLATEAU AND COMPLEXITY

Figure .2.9.2

We can control the dolls above and below because of our digital mental capacities. We can change the natural adaptation/energy flow. By adopting the mechanism detailed in the preceding paragraph (Figure 2.8.2) we act as digital regulators, materialize our ideas, and fix levels of energy flow. Humans can strive to reach goals.

Dolls who are far from the intelligence plateau (Figure 2.9.1) do not strive for goals, this behavior requires digital modeling capacities. Because we have these aptitudes, we can escape the natural analog adaptation system of the Russian doll chain. We create, we materialize new combinations of atoms. These arrangements would not have appeared in the normal evolution of the chain. Our mental space generates extreme ingenuity, persistence, and resilience to materialize abstract ideas. We change our environment and organize it according to a mental plan.

Brain-centrism considers that the intelligence plateau is part of the life plateau. (Figure 2.9.3.) This is because intelligence requires striving for survival, thus curiosity. Without a natural world, intelligence would not appear. That also means that for brain-centrism there cannot be intelligence without evolution. Artificial or mechanical intelligence would be better named advanced automation.

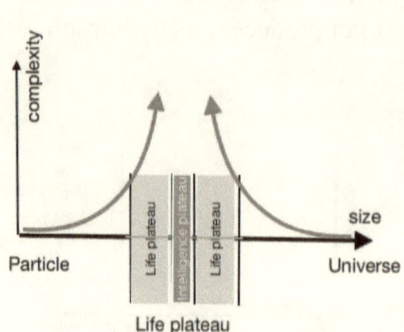

Figure 2.9.3 Life and intelligence plateau

*

Survival of living systems on earth is only possible because of the continuous energy flow from the sun that allows a differentiated inner state.

The *second law of thermodynamics* tells us that everything in the universe tends towards disorder. We can experience daily this tendency at work. Organized systems end up disorganizing and our hot cup of coffee slowly regains the room temperature. This tendency towards thermodynamic equilibrium is a measurable characteristic of a system's representation called *entropy*.

An entity can maintain its high level of organization distant from equilibrium, only with a sufficient inflowing stream of energy. Ilya Prigogine (1917–2003) named such entities open "*dissipative*" systems because they dissipate the incoming energy. They will use this energy to resist drifting towards equilibrium by reorganizing themselves and maintain or increase their complexity.

With a sufficient temperature gradient, dissipative systems *self-organize*. The simplest example of a physical dissipative system that self-organizes would be a pan of water on a fire. The fire maintains a continuous temperature difference between the top and the bottom of the pan. If the temperature gradient is sufficient, one would see cyclic convection currents[19] appear. The system is self-organizing. The system stabilizes far from equilibrium, as long as the energy flow continues.

Typical examples are hurricanes, living organisms, societies, brains, and perhaps even the observable universe. These systems self-organize to maximize the flow of energy. In a living system, this energy will support life and generate additional internal information.

Dissipative and self-organizing systems are the first step; however, we have not found a complete mechanism explaining how the first living cell is generated.

Life appears to be an improbable event. (Figure 2.9.3.) For instance, a protein is made from a ribbon of amino acids. 300^{20} possibilities exist to assemble 20 out of 300 amino acids in the right order to make a human protein of medium size. This huge number makes it unlikely for evolution to stumble onto the correct combination.

Intelligent life in our sense is even much less probable. Maybe, because of our individuations, are we missing important information that would change this probability?

The initial conditions of the Universe seem to be set up for life to appear. We know this problem as the *"fine-tuning problem."* If certain universal constants had been different, according to our models, the chain of Russian dolls would not have developed the same way, and we would not be there to notice it. The *anthropic principle* states that "scientific observation of the universe would not even be possible if the laws of the universe had been incompatible with the development of sentient life."

---

[19] Convection is the transfer of heat due to the bulk movement of molecules within fluids.

In our daily life, we often forget the immensity of where we stand. We also forget that we are in a unique position on the intelligence plateau of complexity.

According to brain-centrism, our stand as part of the chain prevents us to have a complete third-party understanding. We cannot in the verbal world understand the entire chain and its origins because the mental space is here describing itself. This limitation is a fundamental characteristic of the mental space that we call a blind spot. We will examine it in (6.10).

Orders only have a mental existence. They are, as we described earlier, a function of the mental space trying to organize itself. Out there is no order. Orders require mental spaces to individuate them. Entropy, as a function describing order, is thus also a construction of the verbal world. Entropy is useful for the mental world to explain certain perceptions. Out there is no entropy. We are in the middle of the Russian doll chain that our mental space has created. Where else could we center this chain? We are on the intelligence plateau that we have imagined. Out there is no intelligence plateau.

* *

2.10 Extraterrestrial intelligence

Most of us feel the profound impact it would have on us discovering extraterrestrial life. But the impact that would have an extraterrestrial life form would change humanity. Nothing more could teach us about ourselves than such an encounter.

All our knowledge up to now has come from a unique source, the human mental space. Meeting an intelligent alien would show us another mental space at work. Such an encounter, especially if our extraterrestrial is different, would inspire new concepts on "out there" that we cannot develop on our own.

Such an encounter would be as interesting for them as it is for us. I hope that by the time we develop technology for interstellar travel, we will have mastered our fears of Savanna inhabitants!

Most science fiction narratives involve extraterrestrial beings. They describe what they look like, how they perceive reality, how they think, their ambitions, their technologies. However, it's the human mental space that produces science fiction narratives. If we get to meet aliens, they might not compare to any of our expectations. We cannot escape being anthropocentric.

Trying to communicate with an intelligent alien might be a very enriching but challenging task. It will require a complete decentralization of our perspectives. For the first time, we could have a third-person perspective on our mental space.

Would we be able to recognize intelligence in aliens? Do we recognize it in other species on earth? A more practical approach would be to evaluate aliens' productions. If they visit our planet, they can travel in space. If they have come with a spaceship[20], they must have discovered digitalization, language, and master technologies. They are on the intelligence plateau.

There are other types of intelligences, like swarm intelligence or plant intelligence[21]? Could one speak about a swarm's mental space or a plant's mental space? None of this intelligence on our planet has reached a digital level and the intelligence plateau.

We wish to meet an intelligence like ours so we can recognize each other and communicate. We expect they would be for us a true third party.

Would they be organisms or mechanical entities? We would expect them to be organisms. Nature does not produce intelligent mechanisms without transiting through brains. (2.8) Cybernetics tells us they must have a brain and a mental space. If they are organic, they have developed following Darwin's principles. However, their adaptations have been fitting their environment. Humans are "good models" for planet earth; they

---

[20] Certain bacteria can support space conditions.

[21] Monica Gaeliano, an Italian behavioral ecologist wrote: *"Plant cognition is a new and exciting field of research directed at experimentally testing the cognitive abilities of plants, including perception, learning processes, memory and consciousness..."*

would be "good models" for Planet X. If they are mechanical, we would ask what organism has conceived them. Mechanisms need a creator. We suppose organisms don't need one and have developed according to the laws of physics, although we don't know specifically how. (2.11)

Will they come from a planet? Their natural environment must be a planet to produce enough diversity and abundance. A star or outer space would not allow it according to our physics. They would be the leading species on Planet X, but they could not be the unique species. They demand a complete Russian Doll chain, with sufficient complexity on both sides of the chain and quite a long history to have given them the time to reach an intelligent plateau. Their life duration will be long enough for them to have time to learn, but not too long to maintain a renewal cycle.

Would they have a mental space? As we know, they will have an "expectation" system. (1.9) Their mental space will therefore have many similarities with ours. For instance, they would have developed at least one central brain and a variety of external and internal sensors. They would have discovered symbolic languages and mastered digital regulators. Like us, they would navigate between two mental worlds, a natural world, and a verbal world. They would have at least two information sources, and their brain would handle analog and digital information. They could use the first person and third-person perspectives and have difficulties translating their Gödelian information into their language. Their individuation categories might differ from ours. This is the case for various human populations on our planet. Richard Nisbett, in his book: *"The geography of thought,"* shows that Western and Eastern populations have different perspectives. Westerner concentrates more on details, and easterners more on a global vision.

What would be their size? A complex doll in our Russian chain (2.9) can only flourish in a complex environment, according to Ashby's law. If their intelligence is like ours, concentrated in individual brains, their size is an intermediary size comparable to ours.

Would they be able to abstract? They would, without abstracting, it would not be possible to conceive laws describing the regularities out there. And without laws of physics, they could have no advanced technologies. However, their fragmentations of reality "out there" might be different if their sensory systems lead them to different individuations. The way they combine abstraction and generalization might lead them to different visions than ours.

Would they use analogy and causality? We should suppose so, as they would have a mental space with two worlds. One could imagine that their analogies can diverge from ours if their individuations are distinct. Among humans, analogies are very different. What they would call causality may also differ from us. Their logical system may well be non-Aristotelian. It could be, for instance, based on probabilities.

Would they have a motion of space and time? Organisms must be able to observe movement, to survive (1.1). Spotting out movement shows a sense of order and time. Their space will vary from ours, perhaps zoomable like a computer picture.

Professor Edward Hubbard explores questions at the junction of education and neuroscience at the University of Wisconsin. He asserts that mathematical concepts such as Cartesian coordinates or the complex plane, *although they appear by cultural invention, were selected as useful mental tools because they fit well in the pre-existing architecture of our primate cerebral representations.*

Would they die? I guess they would as dying individuals is the greatest way for a species to adapt, to survive, and to develop. Their lifespan may nevertheless differ from ours. But it could not be too short because of the limited speed of light. In an overly short lifespan (milliseconds), they could not achieve maturity and accomplish tasks to be transmitted to the next generation. But it could be long according to our criteria. On earth, experts agree a Bristlecone pine tree (Pinus longaeva) in

California's White Mountain range nicknamed Methuselah is over 4,700 years old.

Would they have religions? As they are organic and have generated sufficient complexity to be on the intelligence plateau, they would be sensitive to the mystery and the unknown. They would have developed invisible worlds and fill-in narratives to overcome the limits of the knowable. They would have then a sense of religiosity.

Would they have mathematics? Once gained a structured symbolic language, they would have created mathematics. Their mathematics would have some similarities with ours, but they could have developed different fields of mathematics corresponding to their intuitions and their physiology.

Astronomers and exobiologists have reflected on the minimal conditions required on Planet X for life to develop. They have also examined the probability of intelligent life to appear. In 2019, the Geneva astrophysicist Michel Mayor received the Nobel prize, along with Didier Queloz, for discovering the first exoplanet in 1995. Since then we have found thousands of planets within a radius of 100 light years. One hundred light years is an enormous distance, but not when compared to the 150,000 light years' diameter of our galaxy alone.

The Kepler satellite monitors 150,000 stars like our sun. Slight periodic dips in starlight is the sign that a planet is obstructing the star's light. Many of the exoplanets orbiting their star at around 150 million kilometers (the distance earth-sun) could have liquid water and could be suitable candidates for harvesting life.

The SETI Institute, founded in 1984, under the initiative of Carl Sagan, is listening to detect any signal that could show intelligence. However, up to now, they have found no sign of intelligent life. As the physicist, Enrico Fermi put it: where are they?

The good model theorem and the Russian Doll metaphor assure us we relate the complexity of an organism to the complexity of its environment. Stephen Gould had already observed that it requires a minimum threshold of complexity in the environment for life even to exist. In his book published in 2015 *Pathways Towards Habitable Planets*, Sebastian Wolf declares:

" ... the property of living systems to possess a high degree of complexity in structure and its response to the environment are discussed since it can be used for remote sensing of extraterrestrial life."

Researchers have defined a "habitable zone" based on the distance between the planet and its star. The habitable zone is the one in which liquid water can exist on the planetary surface. We have reduced this Goldilocks zone to consider the atmospheric gazes that could support life. What we would consider "habitable" for single-celled microbes wouldn't work for complex creatures.

Discovering an intelligent signal would be a cosmic event; it could change our deepest thoughts and dreams.

No other discovery, no other encounter could teach us more about ourselves than non-ourselves, extraterrestrial intelligence.

* *

2.11 Life and Survival are part of Intelligence

The ultimate source of life and intelligence is the expansion of the Universe. The universe is the upper limit of every Russian doll's chain. Therefore, its expansion trickles down and affects all the substructures. We know dissipative systems convert inflowing energy into internal order to maintain it far from equilibrium. (2.9) Order means information. By producing inner order, any complex dissipative system generates additional internal Gödelian information.

In 1922 Alfred Lotka (1880–1949), an American mathematician proposed to interpret natural selection as a fight among organisms for energy. Lotka's principle states that:

Organisms that survive and prosper are those that capture and use energy more than their competitors, biological systems compete for usable power.

This principle would apply between species or even within a species.

We can extend Lotka's principle to thermodynamics of open systems far from equilibrium:

Dissipative systems increase stored energy/information to maximize energy flow.

Self-organization[22] is the mechanism by which we store energy/information. This maximizes the future energy flow potential of the system. Biological systems, dissipate inflowing energy in self-organizing in a way that maximizes future evolutionary potential. Survival and adaptation are the consequences of Lotka's principle. Adapting is a way for the system to maximize future evolutionary and survival potential.

This also applies to the brain who uses the metabolism to maintain its internal organization and generate additional Gödelian information by creating new connections and neurogenesis.

Translated at the level of the mental space, it means the brain's metabolism dissipates into Gödelian information that can favor future energy flow by learning and understanding. These characteristics are very close to what we feel like intelligence.

Lotka's principle could justify why the behavior of dissipative systems is unpredictable. If the system had been predictable, no

---

[22] Self-organization is a physical process that happens in all dissipative systems. Phase transitions are the physical processes of transition between the basic states of matter: solid, liquid, and gas.

additional information would exist.²³ It also narrows the gap between non-life and life.

In his book: *"The True Creator of Everything,"* (2020) Miguel Nicolelis explains how, in the summer of 2015, we were walking on the Montreux Lake Shore. We were once more discussing the origins of life and the nature of intelligence when we stopped to admire the intricate structure of a tree. Our observation led us to express what we called the *third principle of thermodynamics*. Those shapes were not random; they were historical traces produces by energy dissipation. We ignored Lotka's principle from 1922. However, our third principle added an essential twist to Lotka. Beyond explaining survival, it also allowed us to describe in physical terms the tendency towards "intelligence."

The open question was: how does energy dissipate into the various hierarchical levels in the brain? *Spontaneous fluctuations* are neuronal activities in the brain that one cannot correlate with sensory stimulus or conscious activity. They make up most of what's happening but cannot know why. Most neuroscientists call it noise. We interpreted this noise as the trace on the neural network of energy distribution in the brain's hierarchical levels and established a correspondence between neural network activity, oscillation, and the energy flow.

---

²³ Data is informative if it is not previously known

Energy dissipates through the brain's hierarchical levels to ensure a maximal future energy flow.

Learning, organizing our thoughts through analogy and causality, understanding, helps the brain remain far from equilibrium by minimizing its entropy and preparing it for potential future learning. It amazes me to notice that survival closely relates to learning for thermodynamical reasons.

Survival, who has been in chapter one, our fundamental assumption, would then have a thermodynamical origin. If one considers entropy as an increase of *"entanglements,"* survival would then have a quantum physical origin[24].

However, survival not only concerns the individual, it mainly concerns the species. An aging population without limitations on natality would imply an increase in the population. Our planet has limited resources, we estimate it can support 11 billion humans. Whatever this figure is, the day will come where resources will start missing. Nature would then have to arbitrate between two possibilities. Either reduce the average age or reduce natality. Adaptation at the species level and our third law would favor newcomers and reduce aging.

* *

2.12 Curiosity and Learning

Each of our 75 trillion cells has a limited lifespan: colon cells die after three days, skin cells after three weeks, red blood ones after four months, and brain cells can last for a lifetime. Until recently, one believed they could not renew. Joseph Altman (1925–2016) has proven this former dogma wrong in the 1960s. He discovered adult brains could replace damaged neurons. Neuroscience did not recognize Altman's work until the 1990s as it contradicted the prevailing paradigm.

---

[24] Part 2 will cover entanglement and quantum physical interpretation

The theory of learning, which up to then, had embraced the limited view that only synaptic modifications could explain learning, had to be revised. Learning involves deep physical modifications at various levels of the brain, including neurogenesis, new neuronal connections, and synaptic reinforcement. Learning is how the brain renews and reorganizes itself at various adaptation levels. (2.11) We regroup these modifications under the name of "brain plasticity." At rest, for an adult human, we estimate that the brain uses 20% of the body's metabolism. A quarter of which is used to maintain the neurons, the networks, and glial cells, we use the rest for generating Gödelian information by creating representations and memories.

The mental space does not directly perceive this huge brain activity involving over 80 billion neurons, although this work generates it, and keeps it alive and active. Learning and memorizing is instead what we perceive.

Not adapting his representations would mean for the primitive man a rapid death sentence. His strive to survive sharpened his awareness, making him conscious of any unexpected sight, sound, or smell. But it also favored his curiosity and his learning appetite.

Curiosity would come for him at an expensive price. He had to learn the proper balance between the risk of approaching, and the reward of discovering. This problematic situation favored mental space development. A situation where he had to master the fear of risk, imagine all future possibilities, evaluate and study probabilities, invent a variety of escape possibilities, and approach techniques, special movements, and escape routes. Curiosity favored developing both the mental space and the brain in reciprocal interactions.

With language, curiosity, learning, and risk evaluation extended far beyond their basic survival roots.

We ignore how much we have inherited and how much our ancestors had to go through for us to become the humans we are today. We also neglect how powerful is the survival need, and how rich its consequences are for our daily life.

I believe the major reason we prefer a human plane pilot rather than a computer one is that we are sure that he, like us, wants to survive. We can be told that a computer is as performing as a human pilot; the computer still misses the survival component.

Expanding our time horizon has extended adaptation beyond short-term needs, we strive to adapt to future conditions. We not only solve problems to adapt now. We ask questions; we imagine futures, we wish, we hope, we experiment. We want to survive in the future, whatever this future will be.

This is a gigantic step, extending our adaptation levels not only to what we observe but also to what we imagine. We are the only species capable of having visions and dreams about our future. We then plan, agree, and cooperate with others to make things happen beyond the simple necessity of immediate survival. Should we consider then this imagination capacity also as belonging to intelligence?

With language and the verbal world, most human activities migrated from the immediate needs proposed by the natural world to expected future needs. Learning from simple imitation extended to learning narratives. We could know things we had never experienced.

Agriculture which started some seven to ten thousand years ago, for instance, needed a common language to be implemented. It started in the so-called Fertile Crescent, an area between the Syrian Desert to the south and the Anatolian Plateau to the north, where we invented the first written language and the wheel. We know the area as the cradle of civilization.

Agriculture needed knowledge that one cannot learn by experience in a short period. From that moment on, humanity has been striving for purposes well beyond immediate survival. From adapting ourselves to what nature offered, we started with agriculture to adapt nature to our needs by using materialization (Figure 2.8.2).

Since then we have been harnessing the energy and using it not only to survive but also to ensure that we had better survival possibilities in the future. What happened 10,000 years ago in the Fertile Crescent was the first manifestation of a tendency that has brought us to the civilization as we know it today. A migration from analog-to-digital. We recreated our "Reality" with the tools of the verbal mental world.

To end this chapter, and these first remarks on the brain-centric perspective, I want to cite Gottfried Wilhelm Leibniz:

"Although the whole of this life was said to be nothing but a dream, and the visible world nothing but a phantasm, I should call this dream or phantasm real enough, if, using reason well, we were never deceived by it."

Although brain-centrism asserts that our knowledge is mental and our conceptions are not "isomorphic" to out there, it also claims, as does Leibniz, that it can be "real enough" to call it Reality.

* *

Key ideas of chapter 2:

- Human memory is mainly analog and very different from computer memory.

- There are no general neural codes

- The computer's mental space is limited to the verbal world, it has no Gödelian information

- Like expectations and fill-ins, abstraction and order are a mental "motor" we cannot stop them.

- The mental space generates invisible worlds to explain the visible one

- Organisms are complex adaptive systems, they adapt to inside and outside

- Survival is part of intelligence

- Learning is fundamental for neurogenesis and brain connectivity.

-

* * *

III: THE VERBAL WORLD

We have described two worlds in the human mental space and the two types of information. Since language, we have evolved from a reality centered in the natural world to one with all the complexities of continuous interactions between both worlds. The concept of information is central to study the mental space. Information can have several facets. We already shortly described Shannon's digital information and Gödelian analog integrated information. We will examine here limitations of the verbal world implied by structured languages.

> "The ability to perceive or think differently is more important than the knowledge gained." — David Bohm

3.1 Languages and mental spaces

We have distinguished two worlds in the mental space: the ancient natural world and the verbal world. The first is using integrated Gödelian analog information, the second, whose representations are words and narratives, is using digital information. Gödelian information is subtract-dependent, making it difficult to communicate. However, any representation of the verbal world is connected to the feelings and emotions of the natural world.

I can't imagine human reality before language. One can try thinking without words or communicating without a structured language. When we have to do it, we rely on gestures, like pointing to objects or exaggerating facial expressions. If we have to communicate an abstract concept, the exercise becomes somewhat complicated. How can one think of Newton's laws with no language to think with? What remains of our knowledge if we lose the symbols to express it?

The first sentence of the opening chapter of the Gospel of John in the New Testament is:

In the beginning was the Word, and the Word was with God, and the Word was God.

The word would precede everything. Its power makes it divine.

In his book of 1871, *The Descent of Man, and Selection in Relation to Sex*, Charles Darwin (1809–1882) was one of the first thinkers to claim that language was a human invention. According to the book of Genesis, God gave Adam authority to name every being in the Garden of Eden. This would remain the most common belief up to the 18th century's enlightenment and is still very common in the western world.

The development of languages, according to Darwin, happened in parallel to the biological evolution of the body, the CNS, and the brain. It started from animal-like vocalization whose complexity increased. However, being a mental activity, language could evolve much faster than biological adaptations. Naming natural objects has structured the mental space's individuation capacity. It took thousands of years for human groups to agree on how to name what, then how to combine sounds into words; and how to structure words into sentences. These structural combinations had to integrate the beliefs and the cultural rituals of the population, their behaviors, their social hierarchies, and their interactions. It had to mirror their relation to the invisible worlds, as well as their apprehension of natural living and non-living entities. Various types of words and their specific arrangements had to reflect not only cultural aspects but also properties of the mental space.

This evolution proceeded over hundreds of thousands of years.

Although language formed over very long periods, we will simplify by regarding it as an invention that took place 15 to 29 thousand years ago.

Aristoteles considered language so significant that he defines humans as: "creatures who use language." René Descartes (1596–1650), the French mathematician and philosopher, reinforced this interpretation, remarking that we should distinguish human words from animal sounds: *"Humans can think without speaking, but can't speak without thinking."* For Claude Levi-Strauss, language is the basis of culture as: *"We don't inherit it, we have to learn it."*

Let's review how structured languages have remodeled our brain, our mental space, our perceptions, our realities, and human society.

Language and brain structures.

Babies learn languages and after only 2–3 years, they can express what they want. We believe existing genetic conditions support this capability. Babies can hear already three months before birth while still in their mother's womb. They then learn the sound and melody of their mother tongue while the brain develops. Although many brain regions are involved in language processing, neurolinguistic distinguish two major regions. The Broca's area in the left frontal lobe, and the Wernicke's area in the left temporal lobe. The Broca area controls tongue movements and the facial muscles necessary to produce speech. If it gets damaged, the patient can still read and understand spoken language but cannot speak or write. Wernicke's area handles interpretation and meaning. When it is damaged, the person may speak, but his sentences are meaningless. Since the 1990s, modern technology like magnetic resonance imaging (MRI) has produced an enormous shift in neurolinguistics. It now allows us to track the brain activity that is going on while people are reading, listening, and speaking instead of concentrating on aphasia patients.

When we created spoken words, they had to respect our auditory frequency characteristics. Sentences had to account for the capacity of our short-term memory to store and assimilate information. The fovea played a role in writing and the additive structure of the grammar. Individuation influenced our descriptions as sums of separated entities in a sentence.

The brain can learn and generate new representations and new pathways. This has allowed languages to speak about

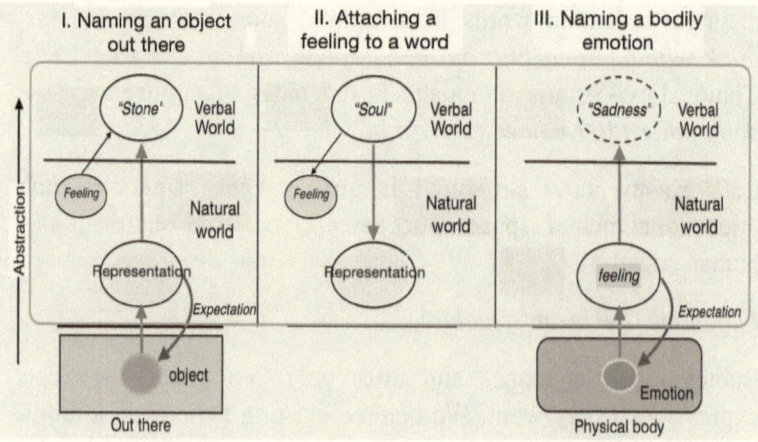

Figure 3.1.1: *I. Naming an object, 2. Attaching a fealing to a name, 3. Naming a feeling*

themselves. For instance, the sentence: this sentence is much too short, is recursive. It has also allowed the possibility to create new concepts as abstractions of existing ones. The brain can create new networks by combining or including elements with existing neuronal ensemble[25].

Language can thus also build nested expressions such as *the table that is in the room at the end of the corridor at the second floor of the house.*

CNS and brain characteristics thus modeled the structure of our languages.

Grammar and syntax.

Syntax is the set of principles by which units of symbolic language (words) are assembled to form more complex expressions (sentences). The possibilities provided by syntax combining concepts into meaningful sentences were so impressive that the usage of symbolic language imposed itself. It offered the opportunity of breaking down complex thoughts in smaller steps and to establish new concepts by gathering

---

[25] The distributed principle, which states that behaviors depend on the coordinated work of populations of neurons distributed across multiple brain structures, describes this brain capacity. In July 2009 Miguel Nicolelis and Mikhail Lebedev a paper in Nature Reviews entitled Principles of neural ensemble physiology underlying the operation of brain-machine interfaces.

existing ones. It made understanding clearer. Syntax allowed us to represent any situation. With an appropriate sequence of words, and following the rules of grammar, one could describe events, objects, or feelings visible, never seen, or invented. Syntax was our initial step towards materialization. (Figure 2.8.2.)

We still debate the question of the origin of grammar. Noam Chomsky argued that grammar is innate and universal. Other linguists, such as Daniel Everett, oppose Chomsky's claims that grammar was invented as a tool to solve social communication problems.

Language gained its life through its freedom of combining concepts, creating extra words, and grammatical categories. In return, these new grammatical categories started imposing themselves on our perceptions and our feelings. We started seeing and feeling according to the language's categories. (Column II in figure 3.1.1.) The architecture of the language reshaped our worldviews, our representations, and feelings while expressing them. The additive structure of the language determined the way we would analyze situations by decomposing them. Descriptive narratives modeled our expectations. We started perceiving and feeling more and more under descriptive narrative and less conforming to our natural representations of "out there."

Narratives and fill-ins.

Narratives penetrated cultures. Certain stories explained how to live a human life, how to raise children, how to bury the dead, or how to organize societies... Other narratives answered existential questions on the origins of the world, on death, and the afterlife. Stories served as fill-ins to understand unexplained things. Most of these fill-in descriptions used beings of the "invisible world" to explain events in the visible one.

Narratives also separated one human group from another one that had chosen other tales as their beliefs. Before language, men could fight for sex or food, they now could start wars for imagined stories. Words became compelling. Their credibility outstripped the one of the sensory system. They produce or mobilize powerful emotions and can galvanize entire populations. Certain individuals became experts in exploiting the power of language.

Language and cognition.

A long-debated question has been: is it possible to think without language? Does language precede cognition or the contrary?

Descartes had explained: *Speech is the sign of thought because to speak is to aim for meaning through signs, and only thought can operate the relationship by which a sign refers to meaning.*

Language and cognition are two functions of the mental space. According to brain-centrism, a thinking process starts in the natural world and thus precedes language that belongs to the verbal world. But things are a bit more complicated.

Ferdinand de Saussure (1857–1913), the father of contemporary linguistics, claimed in his "Cours de linguistique générale" published posthumously in 1916: "*Philosophers and linguists have always agreed to recognize that without the help of signs we would be incapable of distinguishing two ideas in one way. Clear and constant. Taken by itself, thought is like a nebula where nothing is necessarily delimited.*"

For brain-centrism de Saussure is correct, an idea generated in the natural world manifests as analog Gödelian information and is difficult to express before "translating" it in digital form. However, as verbal concepts are high-level abstractions, they never reflect their analog representations source. The verbal abstraction may sometimes trigger, in the verbal world by association, ideas that were not intended. The continuous collaboration of both worlds using different information systems interacting by feed-forward and feedback is a resonance chamber for cognition, who will amplify and multiply thoughts started in the analog world. (Figure 3.1.1.)

THE VERBAL WORLD

The two interacting worlds can belong to two different brains. This is the case when two individuals collaborate in research. One amplifies the ideas emitted by the other.

In The *Relativistic Brain*, Miguel Nicolelis and I described the brain activity corresponding to this interaction between two mental space worlds. According to the relativistic brain theory,

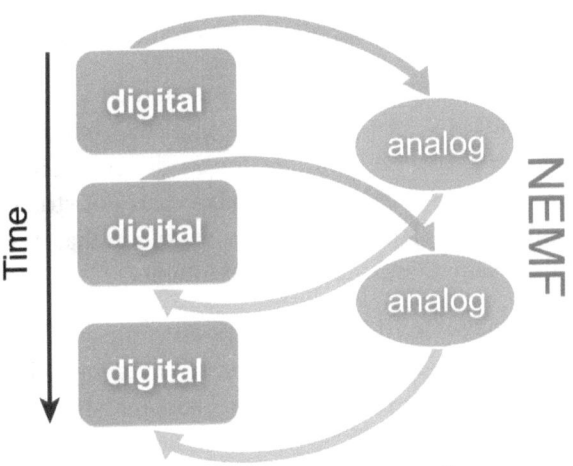

Figure 3.1.2: Hybrid computation

complex central nervous systems like ours generate, process, and store information through the recursive interaction of a hybrid digital-analog computation engine. The digital component of this engine is defined by the spikes produced by neural networks distributed all over the brain, whereas the analog component is represented by the superimposition of time-varying, neuronal electromagnetic fields (NEMFs), generated by the flow of neuronal electrical signals through the multitude of local and distributed loops of white matter that exist in the mammalian brain.

The relativistic brain theory proposes that the interaction of time-varying neuronal NEMFs provides the "physiological glue" creating the neuronal space-time continuum in higher mammals. (Figure 3.1.2.)

Reverting to the mental space description, the hybrid computation engine corresponds to the interaction of the natural and verbal worlds. NEMFs would represent the materialization of the higher brain functions manifested in the natural world, including things like our ability to experience pain sensations, our sense of self, and even our consciousness.

We also attributed the hyper-computation capacities of the brain to the hybrid analog-digital computation engine.

In the 1970s, cognitive linguists maintained a position opposite to de Saussure. They claimed that the building blocks of language—*grammar and lexicon*—depend on pre-existing mental schemas. For them, language does not structure thought; thought shapes language. We now know that both positions combine, natural and verbal world interact.

With language, third-party descriptions transcended our feelings to generate descriptions common to everyone. Individuations stopped being personal choices to become standardized by the lexicon.

Language gave not only shared names to designate objects and events, but it also gave a common way of relating objects to each other and understanding events.

In the natural world, the analogy was the way representations related. An analogy is "personal," each of us makes his ones. The verbal causal relations are shareable and language suggests and pre-embeds them.

Lying and cheating.

For brain-centrism, the most decisive aspect of language is truth. Language gives us the power to manipulate its words. It allows us to build any narrative, with no concern for its truth value. One could lie, express a feeling that is not sensed, or describe a situation that never happened. "Out there" or our natural world cannot lie. Lying is an attribute of the verbal world. It happens because it's a world of symbols. There are no symbols "out there," or in the analog natural world.

Humans understood the advantages of a lie. No experimental verification exists when a narrative concerns non-visible objects or events like feelings, past or distant events, or future ones. Lying is natural for children experimenting with the language it can become disastrous when they turn into adults. A lie needs other lies to cover it up, and the narratives funnel towards complicated contradictions using much of the brain's energy. Many, in human history, have used a speech to build up power or gain trust by making promises and false claims. When nothing distinguishes a true narrative from a fake one, additional problems occupy daily human life. Lies are cancer for societies. In a later chapter, we will consider the imitation fallacy. Imitating, pretending to be what we are not, can also reveal is a dangerous game.

For brain-centrism, there can be no absolute truth as it is the same mental space that asserts the narrative and its truth value. The only replacement we have imagined for "absolute truth" is thus "conventional truth," the agreement between several mental spaces that consider a narrative to be true. (5.3) The scientific method demands, however, that narratives make a falsifiable prediction to be accepted as scientific. It does not mean that the narrative is true, it only means that we have not yet found it false.

Language and Third Party Perspective

Before language, each person was strictly the center of his universe. Everything was "observed" and "thought of" based on a first-party perspective. In the natural world, a third-party perspective is difficult to imagine. This world, dominated by feelings, leaves no space for logical analysis.

One of the most important consequences of language and the verbal world is that it allowed humans to build a "common" verbal picture. In this world, one could describe "reality out there" as independent of any observer, from a "third-person" perspective.

For observable objects, one could be sure that a word referred to the same designated "thing" "out there" for any observer. (Figure 3.1.1.) For entities that one cannot show, a common verbal definition specifying the object through other known words replaces the physical designation. But even if two individual uses the same word, we cannot be sure that they "feel" the same thing. The verbal representation can be similar, but not necessarily the representation and the feelings in the natural world.

Language expressed for the first time a third-party perspective, in which we are not in the center of the picture. The world is "externalized." It exists independently of any observer. Because it is a "shared" symbolic system, language gives an independent "reality" to the event it describes. As everyone speaks about the same "thing," this thing must "exist" per se. The world existed out of us. We perceived it as "being there" even if we were not present, it started having its own "permanence." Over time, this outer existence has become "the reality." Memory evolved, and the explicit component took a growing place. ERP became the predominant worldview.

In third-party descriptions, we attribute qualities to the objects and not to the observer. Let's consider the three sentences:

1.- This red rose is beautiful.

2.- I find the red of this rose beautiful

3.- I find this red color that I attribute to the "thing" we call a rose, beautiful

One and two are first-person perspectives. The first detaches completely "reality" from any observer. The second makes us believe that red is a property of the rose. The third is more correct but heavy to express.

In all three cases, one has individuated and named the rose as separated from the rosier and the rest of the ecosystem. (Figure 3.1.3.)

THE VERBAL WORLD

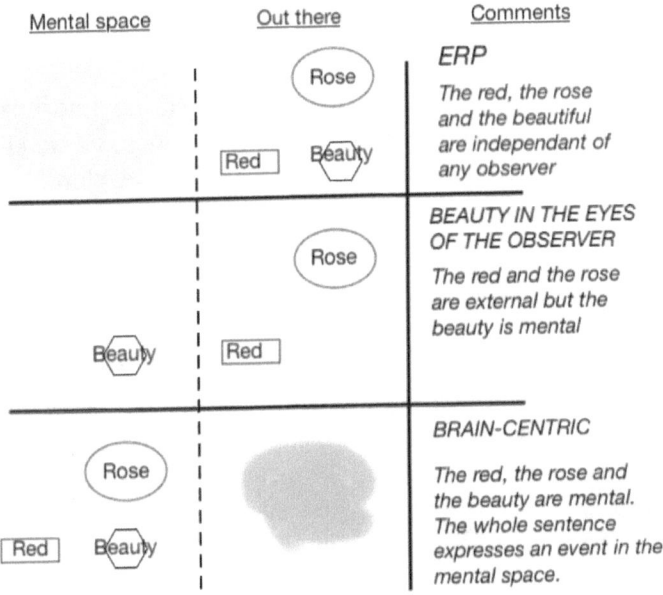

Figure 3.1.3:

For brain-centrism, any description must come from a mental space. What we usually call a third-party description is our best effort to be "objective," to discard any elements in the description that do not come from the object we are describing. For brain-centrism, this is not possible.

Written Language

We invented written language much later, first as analog pictograms, then as symbols: abstractions, representing something other than themselves. Cuneiform, the ancient Sumerian script, emerged 5000 years ago in Mesopotamia, in the Fertile Crescent. It is the first known symbolic system of written communication. Sumerians came up with an amazing and revolutionary idea: instead of representing the visual perception with a pictogram, they started representing the spoken sounds with symbols: the letters. This allowed them to combine sound letters into words, then sentences, just like music. One cannot do that with pictograms. They multiplied this way the full power of syntax and the capacity to write any word that can be pronounced without having to learn new symbols.

Five thousand years ago, humans started writing a new page of their history named digitalization. Digitalization, as we know it today, came over the millennia. It allowed a "third level of adaptation" where it's the environment that is changed to better suit our needs.

Language and interpretation.

Human language is context-driven, not univocal, and not precise. Noam Chomsky declared that its power was in its imprecision. A word can have many meanings, and a sentence is always interpreted. One can use metaphors, irony, humor, hints, or poetry. A sentence can suggest the opposite of what its words are expressing. The imprecision of language allows depiction of the transience of our feelings. Writers and poets can convey subtle emotions and delicate situations by controlling the ambiguities of expressions and sentences. The same word, with a distinctive timbre or a different facial expression, will convey a different message. The language permits humor, the finest expression of human wisdom because it permits lying. Human language is a living system. Anyone can invent a word, and the context may provide understanding. Dialects and jargon can reveal variations or depict meaningful analogies for who can interpret them.

Computer languages.

Computer languages, contrary to human ones, have to be univocal and precise because computers do not interpret. They are not aware of the context and do not use empathy to adapt the meaning of a sentence. As mental space, they only have a verbal world. If computers were subject to interpretation, they would be useless tools, as they would not always give the same result for the same input. The mechanical "reversibility" of computers and their language is their strength and their weakness. The difficulty of computer translation illustrates the confrontation between the two worlds of necessary precision and creative imprecision.

Undefinable Concepts

Some concepts are primitive. Our natural world "knows" them because we experience them. (6.1). However, they are undefinable by description without entering circularities. Among the most fundamental ones, we can name.

Time, space, intelligence, and existence.

Before language, our natural world was already familiar with these notions who are undefinable in the verbal world. After inventing language, human societies needed, however, definitions. We accepted self-referential definitions like time is what a clock measures. We build an instrument to measure time and define time by what the instrument measures.

Circular definitions may be useful in physics, but they can also become confusing and dangerous.

* *

3.2 Maps and territories

In 1974, I came upon an impressive book by Alfred Korzybski, entitled *"Science and Sanity: An Introduction to Non-Aristotelian Systems and General Semantics"* printed in 1933. This title intrigued me and I went through its 600 pages several times. I could not figure out if Korzybski was describing tautological evidence or if he had spotted out a flaw in our knowledge so severe I could not see it. None of my usual discussion partners had heard of him. But 45 years later, I recognize that the significance of Korzybski's work has had an unbroken influence on my thoughts and contributed to my understanding of brain-centrism.

In his book, Korzybski called "maps" approximately what we have named "mental representations." A map is a model of a territory. His major claim is that *"the map is not the territory."*

Figure 3.2.1: Magritte, "This is not a pipe", illustrates that a map is not the territory

That conclusion is not surprising. We know that an abstraction is different "in nature" from what it is about. But Korzybski's slogan reminds us vividly how misleading it is to confuse a "representation" with "what it represents."

Identifying the map with the territory is yet pretty frequent in daily life, but also science. It regularly leads to logical mistakes.

A menu is not a meal. The word "dog" does not bark. The word water is not liquid. A scientific description is not what it describes.

The verbal world is not the natural world, it's about it. The true anchor of our life in the natural world.

You perhaps know the famous painting of Magritte: This is not a pipe! (Figure 3.2.1.)

I still remember the first question that came to my mind:

—If it is not a pipe, then what is it?

We are so deeply immersed in the ERP worldview it makes it difficult for us not to identify map and territory.

The difficulty that Korzybski noticed in most Indo-European languages is the usage of the verb "to be" in the sense of identity. One can say: "This is a book." We identify the object you are holding in your hands with the word book. "Is" is here used as an identity. Korzybski attributes the origin of this misleading construction to Aristotle and his first principle of logic who states, "or something is or it is not." We need something either to be or not to be. If something is not X, it must be something else.

Our habit of identifying maps and territories shows the tight grip ERP has on our thinking process. In chapter one, we have considered perception and noticed that this identification could not be the case. The structure of the language favors this identification. As we use language for any description, it takes a little effort to realize this identification as an abuse.

This confusion can become very problematic in science when the territory is "defined" by the map itself, with no observation, as we will analyze in the next chapters.

The distinction between map and territory supports the idea that our knowledge, our mental models of "out there" differ from out there itself. It alludes to a problem on which René Descartes had stumbled in the first part of the 17th century. How can mental act on physical, if they are not of the same nature?

If one draws a static map of a forest with exquisite precision, it will not show the leaves falling in autumn. We can build a simulation, a "dynamic model." We add the "falling" of the leaves at time t. But "out there" the falling is not "added" it is integral to the system. We have fragmented and by adding the fall we are trying to "stick back together. But whatever the stick backs we make, we cannot build the map the way "out there" works. "Out there" is not fragmented. Sticking back produce mechanical objects, a tree is an organism. Scientific models represent how an individuated object changes with time. They use mathematical differential equations to describe the changes. The models are not "out there." The Newtonian laws of motion, for instance, represent motion "out there," but they are not "out there."

A computer simulation represents what it simulates; it is not what it simulates. If you combine in a simulation representation of atoms of oxygen and hydrogen, it will not produce water and energy but only representations of these elements.

Confusing maps and territory can be very misleading. When one uses mathematical models of "out there," one should remain conscious that we are using abstractions. We are navigating in the mental domain, not in the physical one. Basic concepts that we use daily in physics, such as a point, a line, a circle, a function, equality, continuity, computability, infinity, probability, randomness, force, field have no independent existence in the physical world.

Even within mathematics, the confusion between an object and its representations can lead to misinterpretations. The very appreciated and talented Russian/American mathematician, humanist, and author Edward Frenkel[26] gives an example. Edward underlines, on Numberphiles[27], the confusion between a mathematical object and its representation with numbers. He considers vectors on a plane with an origin point. He then chooses a coordinate system and represents each vector with the two coordinates numbers of its endpoint in this system. The two numbers he explains are not the vector; they represent the vector. The vector precedes the numbers; it existed before anybody described it with numbers. It would still exist if one changed the coordinates system and represent it with two other numbers; its nature is not numbers.

When some thinkers declare the universe is mathematical, we should consider it as a metaphor. Mathematics is a human language used to represent, "out there." Some physicists claim the universe is a giant quantum computer and that physics could

---

[26] Author of Love and Maths, The Heart of Hidden Reality, Basic Books

[27] Numberphiles is an extraordinary youtube channel on mathematics

reduce to quantum information theory. But classical physics, and quantum information theory, are mental representations, they are not what is "out there."

Modern physics has added to the confusion between map and territory. It defines "existence" differently than our commonsense notion of existence. It claims that *something exists if it is useful to explain an observation.* (6.3)

Mental abstractions of the verbal world such as rules, or laws of physics, have no *causal efficiency*; they are made of symbols and do not, themselves, "generate" anything in the physical world. Laws of physics describe regularities we have noticed in our representations of nature out there. Maps are essential to gain understanding, but the causal agencies must be separately specified. Newton's laws of movement move nothing.

"The planets don't know Newton's law, but they still orbit the sun."

Verbal mental abstractions and their combinations remain mental, and not physical.

**

3.3 Information

The word information covers so many subjects that it is hard to say what is not information. In every object or event, we individuate "out there" our mental space can "detect" information.

We only started using the word information in the enlightenment age. It carried the idea that the world "out there" is informing us through our sensory system. Experimentation was the act of collecting "information." Matter "out there" would have a "form" to which we are receptive. Information was thus the mental ability to collect "forms." For the empiricists, giving form meant attributing a unit, like a meter or a kilo, qualifying the measurement. (The idea of forms was from Plato.)

Philosophers, like David Hume, did not agree. He believed that our sensory experiences are much too imprecise to make up a sure link between us and "out there." The question of how the mental space gathers information about the world remained an open question until the discovery of the expectation mechanism.

The word information is self-referral, it can apply to itself as in the sentence: are you informed about the latest information? Because information is self-referral, it can produce paradoxes like: Having no information is having some information. In this sentence, the second usage of the word information is about the first usage. The absurdity results from having the same word representing two different abstraction levels. (Figure 2.4.1.) The second information is "meta-information" and we should not confuse it with "information."

Understanding, intelligence, knowledge, language, secrecy, learning and teaching, are all mental attributes using information. As with any mental abstraction, we cannot see it or touch it. We can give information without losing it, and we can memorize it, forget it, or write it. Information can save our lives; or trigger our curiosity; it can be as helpful as it can be misleading. It can trigger beliefs and destroy empires. It can reassemble people as it can start wars. It can boost the stock market or crash the economy. Information is at the very core of science and measurement, but likewise of communication, justice, and democracy. Producing and sharing it is of major importance at all levels in human societies.

Defining information to cover all its usages and correspond with our internal feeling is elusive. Too many significations of this single word are possible to make it useful as a scientific concept. The word information can apply to anything, anything can be informative.

One can distinguish the capacity of "something" to communicate information from the capacity of mental space to individuate information. Brain-centrism will concentrate on the second approach. ERP focuses on "out there," it will adopt the

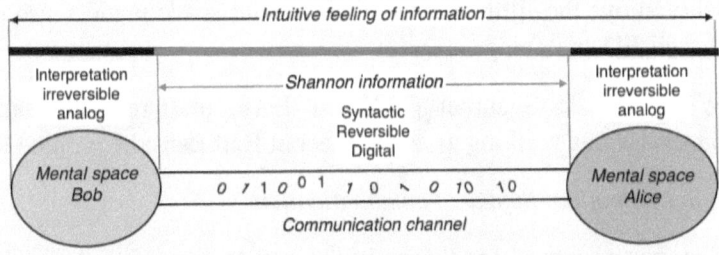

Figure 3.3.1: Shannon information is restricted to the information channel

first. For physicists and computer scientists, information is an intrinsic quality of an object "out there" that we can calculate. ERP will therefore avoid any "mental" interpretation.

The concept originated as a feeling belonging to the natural world, much before the first languages. Later we individuated it and named it. Naming a feeling always implies a problem, only a descriptive definition is possible and it will never reflect all aspects of the feeling.

Science needed a usable definition of information, a definition that would "make objective" the concept, although it is not a measurable property of matter like mass or electrical potential. Aristotle, Leibniz, Boole, and many others had previously opened the road. One had, however, to wait for the groundbreaking work of Claude Shannon (1916–2001) in 1948 to clear it up. Shannon's information theory does not concern the mental space interpretation aspect of information. Any role of the mental space is discarded.

The word "Burro" means butter in Italian and donkey in Spanish. The "objective" part is the five-letter b u r r o, written in this order, with no allusion to meaning. Meaning depends on the language the receiver uses. It does not concern Shannon's information. Remember our James Bond example, burro, viewed as a suite of five letters with no meaning, would be the technician's perspective. He describes switching LEDs. The Italian and the Spanish would adopt a spectator's viewpoint, who enjoys the film; because their mental space made sense of these LED switching.

When Claude Shannon published 1948, his groundbreaking paper, he was working as an engineer at Bell Labs; he entitled it:

A mathematical theory of communication

Shannon proposed his theory in the perspective of studying the maximal information that could transit a communication channel. His information is discreet and digital, the minimum unit is the bit, it corresponds to a yes or no option.

Shannon described his approach to Information as "restricted" to a specific context: sending a message between a transmitter and receiver over a (potentially noisy) channel. The maximum success of this transmission is that the message received is identical to the message sent, with no loss or gain during the transmission. The system is thus reversible, information can be returned to the sender with no loss. It means that the word b u r r o is the same on both sides of the channel, the letters being in the same order.

Shannon's information is thus purely about the syntax. The bits have no intrinsic "meaning." Shannon's concept of information puts aside semantics and interpretation, as it is only concerned by what happens in the transmission channel and not in the mental spaces of the sender or the receiver. Shannon information is digital, subtract independent, and becomes a measurable quantity.

Shannon's information was what ERP science needed. It is subtract independent and thus communicable. It perfectly fits the recently developed electronic digital computers. The success was such that it imposed itself as "the" notion of information. In physics, the maximal information "contained" in an "object" was now a measurable quantity. One just had to evaluate the number of physical states this object could take.

Here is how Shannon comments about his theory of information regarding semantic issues:

"Frequently the messages have meaning; that is they refer to or are correlated according to some system with certain physical or conceptual entities. These semantic aspects of communication are irrelevant to the engineering problem. The significant aspect is that the actual message is one selected from a set of possible messages."

Shannon's information and the languages coded in bits avoid interpretation problems. The receiver has, however, to know how the message was encoded into bits if he wants to understand it. Reading the message does not need a code, but understanding it does.

Here is how the famous Argentinian mathematician Gregory Chaitin (1947-) describes the "cost" of introducing Shannon's Information in "formal systems":

"Formal languages avoid the paradoxes by removing the ambiguities of natural languages. The paradoxes are eliminated, but there is a price. Paradoxical natural languages are evolving open systems. Artificial languages are static closed systems subject to imitative meta-theorems. You avoid the paradoxes, but you are left with a corpse!"

Humans overcome the "paradoxes" of natural language by understanding the context and the mental space of the emitter. They refer to their natural world. For them, meaning is crucial, but they find it through the second information source (Figure 1.6.2), it does not transit through Shannon's information. If we want to account for the role of perception, we will need a broader notion of information. We will then have to accept introducing Chaitin's ambiguities, communication difficulties, and irreversibility. As Chomsky observed, "meaning" emerges in the natural world from analog ambiguities. Defining a word by description does not assure us that all humans will give the same meaning to this succession of alphabetic symbols. The feelings that each of us will relate to this word will differ from one person to the other, making ambiguity unavoidable.

Figure (3.3.1) illustrates the communication between Bob and Alice. Shannon's information concerns what happens in the communication channel, but discards what happens in the two mental spaces.

Robots using formal systems and Shannon digital information do not have these problems. They don't have a natural world. If they did, they would be useless to us. That makes communication with robots somehow difficult. It obliges us to adapt to them, as they cannot adapt to us.

After the 1948 paper, ERP scientists started forgetting about the broader meanings of "information." The quantifiable Shannon information was the information that fit the ERP physicalism paradigm and the only information usable in science. As a result, they excluded considering the mental space. For many scientists and philosophers, mental functions had to be "illusions," because theories do not accommodate their existence.

Shannon had "objectified" the word information by giving it a third-party definition that made a physical sense. Although the information collected is not a property of an object, one can calculate the maximal number of bits of a system by counting all its states.

The language of scientific knowledge, like the popular general language, refers increasingly to numbers. We measure everything to build better models. Numbers and data feed our computers, they are the language computers "speak," but not the language of our mental space, and not the one of our natural world. This drift towards digitalization respects Asby's law, but it mechanizes us.

Shannon's information has invaded our lives and now dominates our relations, our activities, and our worldviews. Personal impressions, feelings, and other "subjective" judgments are discarded for the profit of more reliable, measurable, or countable data. For instance, we now base many tests and exams on multiple-choice answers, which are easier to correct with computers. Algorithmic procedures examine us in the background. Health organizations, insurances, banks, tax offices, are deciding on our behalf. We estimate automatized decisions based on numbers more reliable and cost-effective than personal interviews. The digital age has opened up the doors to a surveillance society. By adapting to the language of computers for the sake of efficiency, we neglect our natural world.

However, discovery in science, creation in arts, novel solutions to human and social problems, all make usage of intuition, feelings, creativity, and imagination. These are attributes of the natural world. They imply the ambiguities of analogy. Discovering something is not an algorithmic activity.

The ERP language and methods used to study "out there" do not adapt to study our mental space. A third-party digital view will "miss" something essential when considering only the brain and not the mental space.

* *

3.4 Shannon information in the mental space

For a mental space to derive meaning out of Shannon's information, it needs to have some previous knowledge. Out of a sequence of bits, one can, "a priori," not know if one should call it noise[28] or valuable information. If we know the series contains relevant information, we still need a "code" to understand it. For instance, if you have a book written in an unknown language, you can, a priori, suppose that the book contains valuable information, but you cannot understand it. You need to know the language or at least have a "Rosetta stone."

Jorge Luis Borges (1899–1986) in his novel, "*The Library of Babel*" illustrates these prior requirements to find useful information. He imagines a huge, but finite library containing all books made of any combination of one million characters. In this library you can find, for instance, all the science books that will exist in one thousand years and all the lottery winning

---

[28] Noise is a random series of bit.

numbers during these years. The library comprises invaluable information. Most books contain unreadable garbage. If one could know what book to read, we would see the future. The library would be a time machine, allowing us to access the future now. There is no way to select the correct book. Even if somebody pretended to have picked the right book, he might have to wait one thousand years to prove it.

We can instruct any computer to print out the library. The library is therefore very compressible, as a small program can generate it. A program that selects the "correct" books and eliminates the garbage is missing. That is also the problem of the internet today.

The mere existence of such a "Borges selection" program would mean that our universe is deterministic: the future would be pre-written. It would also mean that one can solve the "truth problem" (4.4): *is there a way within the verbal world to determine for every narrative if it is true or false.*

The famous twentieth-century French mathematician Armand Borel (1923–2003) has noticed that the number π contains in its decimal development all possible information. We could add an infinite number of Babel's libraries! In Carl Sagan's science fiction novel "Cosmos," extraterrestrial suggests that important information on the creation of the universe lies hidden some place in the decimals of π.

Information is everywhere, the problem is not information, it's selection. Selecting is a mental operation. In 2020 we call it the fake news problem. We realize that our society relies on truth. As most of the information we digest comes from other humans, our entire society depends on trust.

Data is never enough. One needs understanding. One can have a vast collection of data, however, lacking understanding, one cannot sort between garbage and valid information. In (4.8), we will assert that computers will not solve the problem because they have no natural world.

Shannon's information is a mathematical theory belonging to the verbal world. It needs a mental space to individuate it and to extract information from a series of bits. Shannon's information has to be encoded in the matter, with no reading code, matter possesses only "potential Shannon information."

Let's consider that the gravitational pull between a planet and the sun is an exchange of information, one body influencing the other. It cannot be Shannon's information. Shannon's information would require mental spaces at both ends to encode it and decode it. Yet, there are no digital regulators developed by nature. Shannon's information is an abstraction of the mental space. It is about mental representations of "out there," not "out there" itself. At the "ground level" there is no Shannon information. It cannot be an intrinsic physical property of an object. An object "is" itself it is not "about itself." A symbol is not what it represents. Only a mental space can generate Shannon information about some ground-level event.

This is contrary to ERP and the paradigm most scientists use daily and will attract me quite some criticism.

For instance, it's an abuse in the language to say that DNA contains digital information. This digital information only happens in our mental representations of some physical events. DNA molecules are integrated and causal effective Gödelian information. Nobody "out there" has established a code, only mental spaces do that.

Criticisms are now extending to microbiologists.

The transfer of information through perception, and emotions, is physical, it does not use the Shannon digital coded route. Once a physical event has triggered a representation in the natural world, then the mental space can generate further mental abstractions in the verbal world and use Shannon's information.

Shannon's information can be "detached" from the subtract where a mental space has encoded it. It can be re-encoded in other physical subtracts. The recipient can decode it. A computer can manipulate it, and it will allow reversible computation. However, this digitalized information is not causal efficient by itself[29].

* *

3.5 Information in the mental space

We have asserted that information that builds our mental representations comes from two sources. (Figure 1.7.2.) The first source is our emotions. This is Gödelian information, and emotions appear as feelings in the mental space. Gödelian information is integrated, not symbolic. When you feel pain, it's not "about" pain, it's pain itself. If you feel love, it's not about love, and when you fear, it's not about fear. We do not translate emotions into symbols in the natural world. They are physical. They existed millions of years before any symbolic representation.

We feel pleasure or dislike directly; we need no word for us to know about them. It's "direct knowledge" that we only gain through a first-person perspective. This direct knowledge has developed to make the brain aware of inner events. We experience it, we feel it. At a later stage, we can then describe this analog information with words.

Discussing with Miguel in 2014, we agreed that this "direct knowledge" is also information physically active in the brain. We should not neglect it when studying the mental space. We were observing at that moment a tree trunk on the Geneva Lake Shore, and we could gather a quantity of information stored physically in its matter. The convolutions, the shapes, are significant traces.

A first-party feeling is not "independent" from its subtract, it is its subtract. That is why when we translate a feeling into words,

---

[29] This lack of causal efficiency of Shannon's information was the basis of Descartes mind and body problem. How can the mind act on the body? (3.10)

we cannot be precise. Natural language gives us the possibility to use words poetically, to help up overcome this difficulty.

One can conjecture that the physical, integrated Gödelian information is present everywhere out there. We can see it at work in all the physical forces. When a stone is falling, for instance, it exchanges Gödelian information with the planet. We expressed that by a metaphor:

Planets orbit the sun without knowing Newton's laws.

Because the brain is on the "complexity plateau" (Figure 2.8.2) it has feelings and can express the first-person viewpoint in a communicable way. We thus have access to Gödelian information by feeling. Entities of lesser complexity do not.

A third-party perspective, expressed in the verbal world through Shannon's information, does not account for the enormous amount of Gödelian information in the natural world. It is a fragmented abstraction.

Therefore, to describe the mental space, Shannon's information will not be sufficient. We cannot appreciate certain major aspects of the mental space's activity through a "third-party" perspective. The internal first-person perspective, modulated by physical Gödelian information, paints a different and much richer reality. In most of its aspects, it is not interpretable through the concepts, individuations, and languages used in a third-party perspective.

For planets and objects with low internal complexity, it suffices to describe them in the fragmented ERP way object under the influence of external forces. Third-party descriptions like Newton's laws are then adequate to predict. When we study an organism, a mechanical description will fail. The additive structure of language does not qualify for systems that have developed and increased complexity.

Paul Davies gives a striking example. If you throw a dead bird you can predict its trajectory, but if you through a living bird you cannot.

For organisms on the complexity plateau, with multiple levels of adaptations and feedback like the human brain, third-party models cannot carry out predictions. Lotka's principle explains this lack of predictability. (2.11)

When you evoke a souvenir, you experience sensations, smells, tastes, pictures you don't experience data. The events that make your life don't come to you through measurable properties. They have no weight, no size, no electrical charge. Your struggles, your memories, your beliefs, your understanding, your feelings, your fears, the deepest part of what you are, express themselves in Gödelian information. The words you write, all you assert in Shannon information, constitute only a thin add-on layer and hardly reflect the multidimensionality of your inner life.

We live our life in Gödelian information, but we communicate about it in Shannon's information. Understanding means associating verbal representations with feelings and images in Gödelian information. The natural world gives meaning to the verbal one.

There is a vast difference between experiencing a Beethoven sonata played on the piano (the analog version) and the same sonata encoded with musical notes on partition paper (digital version). Both versions contain identical Shannon information, but the second provides us with a richer experience.

We mentioned in a footnote that many of us believe we have access to the third source of information. One could call it a spiritual source. From our shaman to Descartes and Russell or Einstein, many thinkers have expressed this idea that they considered as a direct relation with God or with the Universe. This third information source would also use Gödelian information and is difficult to translate into words. This third source also generates feelings, sometimes called revelation or inspiration. Srinivasa Ramanujan (1887–1920), one of the most astonishing mathematical geniuses of all times, claimed that he received in the visions of his dreams of entire rolls of complex mathematical formulas given to him by the Goddess Mahalakshmi.

Figure 3.5.1 Gödelian and Shannon Information

Gödelian information	Shannon information
physical, analog, integrated	symbolic, digital, "about"
shaped by natural selection	shaped by grammar logic
limited communication mainly by empathy	symbolic language social communication
analog	digital
feelings	words
inprecise	precise
ancestral	recent
understanding, belief, pleasure, fear, hunger, intuition, creativity, fast, artistic expression	reasoning, cheating, lying, making hypothesis and plans, predicting, calculating, solving, materialization
enactive and iconic knowledge	symbolic knowledge
physical fill in	verbal fill in
little abstraction capacity	unlimited abstraction capacity
learning by acquaintance	learning by description
causal effective	non causal effective
dominating in the natural world	dominating in the verbal world

Panpsychism is the view that the "mind" is a fundamental feature of "out there" which permeates the universe. Thales, Plato, Spinoza, Leibniz, William James, Alfred North Whitehead, all had a similar view. The mind would then come to us through this third channel. Bertrand Russell's Panpsychism, called neutral monism, is the belief that "all causal properties are extrinsic manifestations of identical intrinsic properties." This belief is thus compatible with the universal Gödelian information hypothesis. Russell further believed the "conscious mind" resulted from such a physical structure.

Brain-centrism can accept the idea of a third information source, but as it is also Gödelian, one can only distinguish it from emotions from the first-person perspective.

Computers only use Shannon information, they can imitate feelings or understanding, but they don't live them. You can program a computer to "imitate" crying when he listens to Mozart, and he would do it very well. You can also switch the program so that he laughs by listening to the same music. Imitation of feelings is not feelings. (7.1)

* *

3.6 Learning

We divided humans' mental space into two zones, the natural world and the verbal world. Learning means building mental models and logical context representation networks, Net(R)'s. However, these models are different in each world. Learning in the verbal world is a digital activity. We learn symbols for things and connect these symbols by causality to build narratives. Net(R)'s that our mental space structures extend to both worlds. The new narratives should be compatible with the ones we have learned as Net(R)'s include rudimentary logical principles.

Learning in the natural world is an analog association process. We experience or feel "things" (not symbols). We memorize pictures, sounds, smells, ... These images call one another by analogy. Causality and logic are not involved. The process is physical, we learn by doing.

We can learn about the Gizeh Pyramids by reading a book, or we can visit them. The feeling and the knowledge are different.

Bertrand Russell distinguishes two methods to gain knowledge: knowledge by description and knowledge by acquaintance.

We learn by description by listening or reading a verbal description. This symbolic knowledge belongs to the verbal world. That is the way we learn history, mathematics, ... it's the way schools do their teaching for most subjects.

We get knowledge by acquaintance, by direct experience, by interacting with the subject we are trying to know. We can learn by acquaintance matters who are not accessible by description, such as playing ping pong, playing golf, riding a bike, dating a partner, appreciating music, how to walk, what the red color is, the smells of a rose, what it is like to like or dislike, the taste of an orange, truth, fairness, dignity, kindness... When we learn by an acquaintance, we are forming integrated Gödelian information, with no intermediary symbols.

The pathways for description and acquisition are different, although both use brain plasticity and neurogenesis.

A baby can only get his original knowledge by acquaintance. Our ancestors, before language, could likewise only learn by an acquaintance.

We forget most of what we get by description, five years later less than one percent will remain. But what we get by an acquaintance, we will never forget.

Imitation is an acquaintance. We are not solitary creatures, disconnected from one another. Learning about oneself is workable because we engage with others.

We base any knowledge on a description of previous detached knowledge by acquaintance. This is how my friend the philosopher Craig Weinberg puts it:

"We cannot describe numbers, or gravity, or matter to a creature that does not experience them either... Things which are presumed to exist independently of subjectivity, such as matter, energy, time, space, and information are themselves concepts derived from intersubjective consensus."

Bill Gates wrote in 1993, after visiting central Africa with Melinda.

While I was aware of the vast inequalities in the world, I had never come face to face with people living on the bottom rungs of the income ladder—people so poor they didn't have shoes. Of course, they lacked much more than that: healthcare, education, better seeds and tools to grow more food, and, above all, opportunity. Seeing this with our eyes was life changing.

Experiencing offers something specific that description cannot propose. We connected experiencing to emotions as we express it in Gödelian information. It makes it more convincing and easier to memorize.

Alfred Korzybski, we already cited, had observed in 1930 that there are two types of definition we can give to a word:

An *Intensional definition* is defining a word by associating it or equating it to other words, like does a dictionary. Korzybski's favorite example is Plato's definition of a man. Although it may be correct, it says nothing about the man himself: "A man is a featherless biped."

An intensional definition creates associations between words, connections of mental representations in the verbal world. It presupposes you know all the words used in the descriptive narrative. This may lead to circularities, like in any dictionary.

Extensional definition. This is how you would teach the word "man" extensionally:

While you point your hand at a series of men, one after the other, you pronounce each time the word "man." You rely on the brain of the student to associate what he sees, pointed by your finger, to what he hears. Here, the association is between representations of different worlds: in the natural world, the visual one you pointed at, in the verbal world, the word you pronounced. The idea of repeating the gesture and designate different men allows the abstraction to cover many types of men and learns to neglect the unimportant differences in the abstracting process.

We cannot show mental objects like the Big Bang. We have no other solution than to describe them intensionally. To simplify the definition, we often associate a metaphor or find an evocative name for the "non-observable" object. The name or the metaphor then guides the feeling one associates with the concept. A terrible choice can be very misleading and drive us to a false understanding and critical decisions. In the Big Bang, for instance, nothing banged, and it was all but big! *Naming something can be a disaster for generations*, said the French philosopher Albert Camus.

A careful reader will find dubious names in most branches of human activity.

A computer only "learns[30]" by description as its program is based on symbolic digital language, and his memory is digital. Recently, deep learning system is frequently cited. The program "learns" to relate a word with many unique pictures showing similar objects in various positions, colors, backgrounds, at different levels. The computer makes up a multilevel database of pictures associated with this word, allowing him to recognize it. That is not the way the human brain recognizes situations, patterns, and objects.

The power of a digital language is precisely that it is symbolic; one can learn the symbol without knowing the meaning. Symbolic language is "detachable" information that one can transfer from its original subtract into a new device.

In the last 15,000 years, we have shifted our knowledge from acquaintance to description. Emotions, that our mental space previously only associated with perceptions, we now also associate them with words. We can be afraid of the word lion and not only by the perception of a lion. Language has made it possible for us to hide behind words. We can appear to be, even if we are not. (7.1)

---

[30] In fact he never learns in the human sense as learning implies attaching an emotion.

We now base some of our new emotional landscapes, hopes, perspectives, and fears only on narratives, for which we have no direct knowledge. We trust.

This behavior has changed not only the mental space but also the physiology of the human brain. It is now "adapting" to technology, it has itself created as its new environment. This is changing our habits, our behaviors, our feelings, and developing our beliefs.

Institutionalized religions had imposed authoritative knowledge. The enlightenment had put reason and empiricism at the steering wheel for truth. But authoritative knowledge is making a great comeback with broken democracies, dictators, and sadly often with organized "science" itself.

Solutions we choose to adopt in personal and social life adapt to the knowledge we have, as do the dreams we dream, and the future for which we strive. By considering that the only "real" knowledge is the one that comes from digital descriptions, we possibly forget some aspects that have contributed to our survival up to now.

Mental abstractions and digital language have opened up another drive for evolution.

We imagine a desirable future and try to materialize it, instead of allowing for random mutations to determine our future.

Natural selection has allowed certain species to survive for hundreds of million years because selection and adaptation mediated the random push to the environment. It has proven to be a useful guide for survival. Will, our mental adaptation to "materialized mental abstractions" reach the same performance?

We used to adapt to an environment that developed independently.

Today we adapt to an environment that we create ourselves. We have put ourselves under control of our economy, presupposing that some "invisible" hand knows where we have to go!

3.7 Reduction and Inclusion

A deck of cards is an abstract entity of similar components, the cards. The abstract concept deck focuses on the analogies between the cards. It neglects their differences, such as their color and their denomination. It also discards meta-information on the cards such as their order. Our mental space permits this abstraction by individuating a virtual object, the deck. When you consider a fork, you individuate an abstract entity made of similar metal molecules in a specific order, the form of the fork, and you neglect differences such as the position and the momentum of each molecule. (Figure 3.7.1 first column)

Reciprocally, instead of abstracting the deck from the cards, one can decompose the deck into parts, its cards. The figure represents abstracting and decomposing by arrows in the opposite direction (Figure 3.7.1).

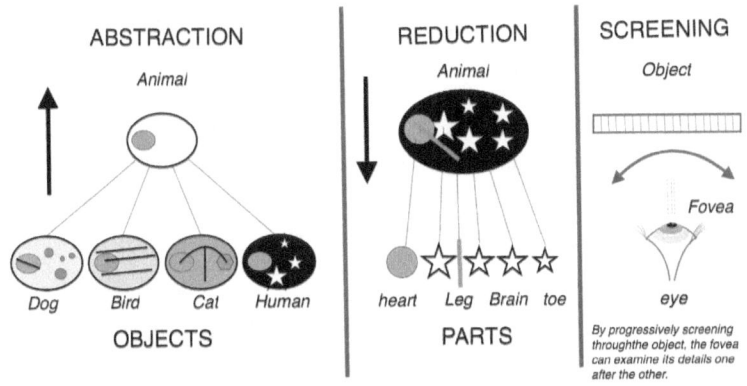

Figure 3.7.1: Abstraction, reduction Fovea

We decompose when we ask: what is this object is "made of"? This question is the method we use when we inquire, we collect information, by fragmenting and individuating parts. We then "re-glue" the parts together by making a sentence. Knowing, in the verbal world, requires decomposing an entity into smaller parts. We then consider that our object is a "sum" of these individuated parts. We call this knowledge strategy reductionism. It is the only strategy used by the verbal world. However, it is not the only one in our mental space.

Often, reductionism suffices to answer our questions; in other cases, it is not. Let's say you decide to sort your deck of cards by color. At the abstract level of the deck, nothing changes, it remains the same deck, but in a different order.

Abstracting builds a higher-level verbal representation by focusing on similarity and neglecting differences at lower levels. A reduction is the reversed operation, it decomposes an object into its parts. (Figure 3.7.1.)

Our mental space can only individuate an object because this object has a boundary. There is "something" larger outside in which contains it. We cannot even conceive something that is not a part of something else. We sometimes call this basic ancestral character of our mental space the inclusion principle. It states that any mental representation "S" in the brain is part of a larger representation. Our mental space always considers there is a larger representation containing "S." The inclusion principle is the opposite of reduction. Inclusion and reduction are the two directions in our Russian doll metaphor. The limits are always problematic if one considers that our abstract representation corresponds to something "out there."

The fovea that we met in (3.1) is a bit in the retina's macula (about 1% or the entire surface) that provides the clearest vision of all. It handles the sharp central vision that is necessary for primates to make a detailed observation. The fovea mobilizes over 50% of the visual cortex. It developed with perception and brain mapping systems. As a theater projector, the fovea lets us follow the exciting part of an event by sweeping through it.

The eye fovea guides the brain's awareness when doing a detailed observation of an object. We consider one region of the object after the other, following the sweeping movement of the fovea over the object to discover its details. This visual characteristic has led us to fragmentation and reductionism, which we now consider so "natural" and logical. It finds its origin in the physiology of our brain rather than in nature "out there." Reductionism thus analyses and describes a phenomenon in terms of the individuated constituents. We can decompose these parts into subparts. This way we can build mental layers of description at increasing resolution levels.

This process is recursive and has "in principle" no end for the mental space who can imagine zooming in. For physics, that describes observable entities, the recursive process needs to stop. Physicists stop this recursivity when the next level of sub-elements can have no more measurable influence on the object they are describing.

Recursiveness is an important property of the verbal world caused by individuation, fragmentation, and reductionism. The choice of physicists to stop it is legitimate. The discovery by Max Planck (1858–1947) and Albert Einstein (1879–1955), beginning of the last century, that electromagnetic interactions only appear in discrete quantity comforts this choice. (5.12)

Reductionism produces convincing descriptive models when the object studied shows negligible internal and external interactions. We then considered the object as "isolated" from its inner parts and the environment, justifying its individuation. Isolation is an approximation, out there, no object is isolated. But even for isolated objects, reductionism will not always give a proper picture of the whole, as in the following example (Figure 3.7.2).

Consider the two-dimensional surface that mathematicians call the Moebius strip.

Its main observable characteristic is that it only has one side. One can draw a continuous closed line on the strip that will mark both sides of the paper with which we made the Moebius. If we now divide the strip into small regions like on the figure and study these regions one by one, none of them has the characteristic of being one-sided. The reductionist analysis, region per region, will miss the key feature of the Moebius strip, which is a property of the whole not reflected in the parts. (See also 3.8.)

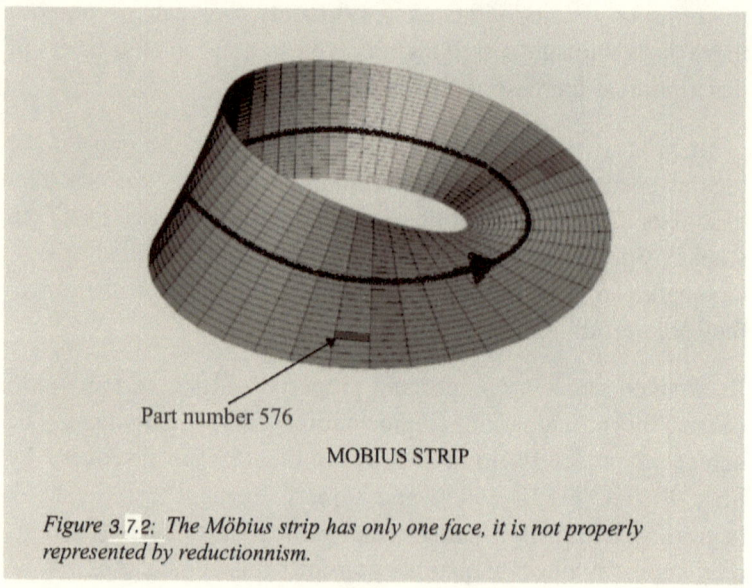

Figure 3.7.2: The Möbius strip has only one face, it is not properly represented by reductionnism.

Often, some physical characteristics only appear when one considers a holistic view, as in our example. This is true for complex interactive systems, like organisms. Up and down the ladder of resolution levels, specific properties appear or disappear. "Out there" cannot be the cause. "Out there" nothing changes when we change our observation resolution. Changes resulting from the mental space's functioning, its abstractions, its individuation, and its fragmentations.

However, our human verbal world cannot apprehend "out there" differently. Reductionism, like individuation, does not bias such as we described in (1.3). No "corrective" methods exist. The "understanding" that results from reductionism is proper to our mental space, it is our mental construction and cannot give us a faithful picture of what is out there.

In our reductionist descriptions, physical systems are reducible to particles. Any phenomena should then be understandable by analyzing subatomic particles. It is not the case. We need other description languages when navigating up and down the resolution ladder. Every level has its language and brings its specificities: chemistry, biology, psychology, sociology. In our deck of cards example, after reordering the cards, it is always the same deck, but when one looks closer, something has changed. By looking at one card at a time, we cannot understand the arrangement order.

Lee Smolin (1955-), one of the brightest living theoretical physicists, proposes an interesting rewriting of physics:

"Rather than describing an isolated system in terms of things that are measured from the outside, we're taking the universe as comprised relations among events. The idea is to try to reformulate physics in terms of these views from the inside, what it looks like from inside the universe."

To account for interactions between parts, or between an object and its surroundings, physics considers another mental abstraction: the physical laws. Our new maps, when complemented with these laws, describe the time evolution of a system that we call models.

On the map, we have fragmented "object" and "interactions." On the territory, the interactions are not "added," they are indissociable from the things themselves and their environment; they are integrated. We cannot observe "interactions"; we only can perceive what they do. When we measure the intensity of an electrical current, we measure what this non-observable current does to our instrument. When you measure your weight, you only read what the weight of your body does to a scale. On the map, we have two separate "things"; the objects and the interactions. On the territory out there "things," are indissociable.

"Out there" physical actions do not use computation, or a representation system to move, there are no fragmented individuations. This is the idea we have summarized by saying that:

Planets orbit the sun but ignore Newton's laws

One could say the same thing for all physical actions:

The needle of a compass will move according to a magnetic field without calculating Maxwell's equations.

Movements on the territory are not "computed," no Shannon's information can be involved in the process "out there." It only gets involved when our mental space models it with fragmented representations. Our models are sometimes "good" enough, but they can never be isomorphic. No computer simulation can model a natural phenomenon.

Henri Poincaré (1854–1912), to account for internal interactions in our models, has added higher derivative terms to the descriptive differential equations. These higher derivative terms can make the equations non-integrable.

Organisms, living systems, are in perpetual exchange with their environment and adapt by changing themselves at various levels, from the molecules to the whole organism. (2.9) There is no way to consider them as isolated systems and adapt reductionist methods. A brain is all but an isolated system. We saw in (2.8) that it belongs to the complexity plateau. These characteristics make it impossible to model it. A global theory of the brain will thus not be mathematical, neither will it emerge from a reductionist approach.

If one considers just one level, the proteins, we estimate that at least 16,000 different ones populate the brain. They formed these proteins as sequences of amino acids threaded like pearls on a necklace. The DNA gives the instructions to position the amino acids. Once the pearls are in the correct sequence, the protein folds up in a precise three-dimensional structure that determines its role in the organism. The folded protein shape represents the lowest free-energy state. This ribbon folds itself up in very complex ways. The geometry of the folding determines what the protein will do. Folding takes less than a second, even if the protein contains several thousand amino acids. A computer simulation, even on the fastest supercomputers, could take 10^{127} years to determine the folded form even for a small protein of 100 amino acids. Recently (2021) new algorithms seem to have significantly reduced the calculated folding time to a few days.

However "out there" is not doing the same kind of algorithmic computation to perform its folding! (2.8)

Even if maps and territory arrive at the same result, they are not taking the same route and not using the same tools to reach this result! Identifying any model to what it models makes little sense, identifying map and territory, as ERP suggests, can lead to terrible confusions. It's true as the system gets more complex with more levels of adaptation to consider.

Digital computing has trained us to think in terms of "detachable" Shannon information: hardware and software. "Out there," this distinction does not exist. Falling apples don't follow any software; they don't need a language; proteins fold using no software to do so, a hurricane does not follow software to take their shape. Only our mathematical verbal world representations of these events require this separation.

* *

3.8 Complex adaptive systems

We have already described why the verbal world has to consider the abstract concept of *complex adaptive systems* (CAS). (2.9) These constructions of the verbal world are necessary to describe certain systems out there. Because we understand a system by answering the question: *what is it made of,* we decompose it in part by reduction. (3.7) If these parts are themselves active agents, re-gluing them to get the complete system is complicated because the parts are not isolated systems.

Imagine a jigsaw puzzle made of pieces, but instead of having a fixed shape, the pieces change because they interact. Re-gluing by adding the pieces will be complicated. In our Russian doll example (2.9), information flows up and down the system. The complexity grows when we approach the intelligence plateau because the subsystems handle information differently. (2.11) They use this energy flow to reorganize themselves. This reorganization generates additional information, like our jigsaw puzzle pieces that would change their shape.

Out there systems are integrated they are not fragmented, our mental space fragments them to understand them. Out there, there are no complex adaptive systems. CAS are on the map, not on the territory as ERP considers.

The mental space that generates complex adaptive systems must consider itself as a system. So is the underlying brain. CAS's are powerful mental tools, authorizing us to examine characteristics of our verbal world and organize the data our instruments collect. It's an abuse of language to speak about a complex adaptive system. We should rather than say the complex adaptive mental representation of a system.

For ERP, complex adaptive systems model open "physical" systems. The components of these systems interact, they adapt to each other and change the entire system. We cannot describe them as the aggregation of static components. The continuous multilevel adaptations imply they change in time. A stream of inflowing energy keeps them far from equilibrium and feeds the changes from one level to the next. Feed forward and feedback combine at every level.

Here are some examples. Living organisms, crowds, flocks of birds, traffic jams, the economy, the weather, the market, companies, human society, the solar system, galaxies. Most of the systems we live with are complex adaptive with multiple layers of adaptation. That makes our simple school mathematics inapplicable in most ordinary situations.

Complex adaptive systems exhibit properties we cannot observe/deduce by analyzing the components, we call them "*emergent properties.*[31]"

Is the "one-sided" property of our Moebius strip an emergent property if we consider it built out of the small rectangles as in (Figure 3.7.2)? The strip is not a complex system, and the property of being "one-sided" is not emergent. The rectangles are not open active agents, there is no energy flow. Looking closer, our rectangles are not rectangles, they are distorted.

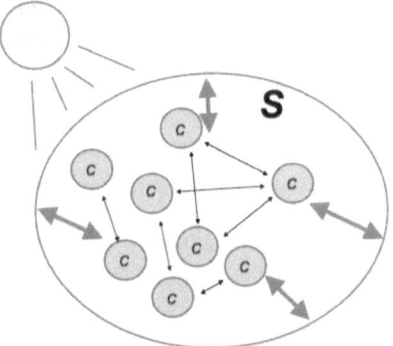

Inflowing energy will generate order in S and generate new information as emergent property.

Figure 3.8.1

Let's take a simple system with two levels, S would be the entire system, the higher level, and $c_1 \ldots c_n$ its components, the lower level. S and c's are open entities. (Figure 3.8.1.)

Let's say the observer is describing c's. Energy is flowing through S, who transforms part of this energy into order or additional information, as we have seen in (2.9). It does that by organizing its components, the c's. In return, c's also interact absorbing order from S and thus changing S. Nothing in the language and the concepts describing individual c's will allow to expect what happens with S. From the c's point of view, a change of S's behavior is an emerging property.

For instance, we cannot deduce the activity of a formicary by observing a single ant. It's an emergent property. The market cannot be estimated by observing only one stock.

---

[31] It is also abusive to say emergent property, it leaves the impression that the emergence comes from the system when in fact it comes from the representation in mental space.

One cannot describe emergent properties with the language, and the individuations used for its subcomponents. They are not at the same level in the abstraction tower. Analyzing a molecule of water will not tell us if the water is liquid, solid, or gas. Liquid solid or gas is another mental individuation that does not belong to the language we use for a single molecule.

The philosopher David Chalmers (1966 -) distinguishes weak and strong emergences. He defines weak emergence in ERP terms:

"We can say that a high-level phenomenon is weakly emergent with respect to a low-level domain when the high-level phenomenon arises from the low-level domain, but truths concerning that phenomenon are unexpected given the principles governing the low-level domain."

One can notice that Chalmers's definition relies on what the mental space expects should happen. Without a mental space that expects something to happen or not, there is no emergence. With weak emergence, we cannot have access to the information that would let us calculate the emergence.

Some usual examples of weak emergence could be :

When unexpected complex patterns form in the cellular automaton, like in Conway's Game of Life[32]. In traffic jams. In the behavior of a school of fishes. In solar systems or galaxies. The emergence of phase changes.

Understanding these emergences requires further levels of explanations. Our initial individuations of the system and its parameters omit some information that becomes important after some time.

Often the omitted parameters were impossible to measure, like, in traffic jams, the reaction moment of each driver. Other times, the omitted parameters will interfere only at a moment, like in phase changes that happen at a temperature.

---

[32] https://en.wikipedia.org/wiki/Conway%27s_Game_of_Life

Chalmers defines strong emergence in ERP terms by:

"We can say that a high-level phenomenon is strongly emergent with respect to a low-level domain when the high-level phenomenon arises from the low-level domain, but truths concerning that phenomenon are not deducible even in principle from truths in the low-level domain."

In ERP, the structure of the universe should cause this "non-deductibility." By having in principle no cause, the emerging phenomenon appears then like a miracle. As Einstein had claimed, we must be able to understand the universe. Everything happens because of a cause, even if at the moment we don't know it.

Can we observe "out there" examples of strong emergence? Such an observation would be the appearance of "something" with no cause. This something would not only be unexplained, but we could prove it unexplainable. This would contradict physics and its conservation laws.

Most neuroscientists consider consciousness and our mental space as "emerging" from the activity of the brain. ERP would require this emergence to be weak. That is why, for them, there must be a "brain code." This brain code would be a way to explain the content of what we experience (the emergence) based on the brain's electrochemical activity. Experience and consciousness would then only be a "weak emergence," and our incapacity to expect the emergence would no more be "in principle."

For ERP physicists, it should be possible to explain consciousness and the mental space with new principles, but within the framework of physics. The mathematician and Nobel prize in physics Roger Penrose (1931 -), for instance, believes that "new physics" is necessary to understand the emergence of consciousness. He developed a theory with the American anesthesiologist and professor at the University of Arizona Stuart Hameroff, in which the brain uses quantum phenomena in the cell's microtubules for its functioning. However, new physics must "fit in" the third-party language of physics. That condemns it as an explanatory system for consciousness and the content of the mental space.

In the verbal world, proposals can "emerge" (be formulated), the truth of which we cannot, even in principle, deduces from truths of elements within the system. The truth of these proposals is undetermined, they have no causes[33].

For brain-centrism, mathematical indeterminacy proves that strong emergence exists in the mental space. We encounter indeterminacy when the mental space examines itself in a self-referral process. (6.10) Consciousness can feel itself in the natural world. When it wants to explain itself through a third-party perspective in the verbal world, it encounters a blind spot. (2.11)

** **

3.9 Causality

Emergent properties are surprising because they seem to escape causal explanations. Brain-centrism considers causality as a function of the verbal world. Its function is to organize, connect concepts, and facilitate memorization. The underlying brain generates it in its effort to minimize energy expenditure, reduce entropy, and increase connectivity.

ERP, on the contrary, would consider that the universe is causal. Scientists keep scrutinizing it to find causes explaining the phenomenon they observe. Physicalism explains the physical nature of causality by energy exchange between two events.

Many philosophers and scientists have rejected the idea of "physical" causality, considering that causation was only a mental operation. For instance,

---

[33] Kurt Gödel proved (4.6) that this can happen for any consistent mathematical formal system. In (6.10) we will analyze why this happens in the mental space.

David Hume (1711–1776) the great Scottish Enlightenment philosopher, argued that there was no causality in nature, although he recognized causality was useful in everyday life. *"Causal necessity has neither an empirical nor a logical foundation."*

Bertrand Russell concluded that, since cause and effect play no fundamental role in physics, "causality" should be removed from the philosophical vocabulary altogether. *"The law of causality, like much that passes muster among philosophers, is a relic of a bygone age, surviving, like the monarchy, only because it is erroneously supposed not to do harm."*

Carlo Rovelli (1956-), a famous Italian physicist, and author writes: *"If I observe the microscopic state of things, then the difference between past and future vanishes ... in the elementary grammar of things, there is no distinction between cause and effect."*

Most commonly for physicists, causality formalizes the statement that information propagates from present to future. However, Einstein's Special Relativity has, in 1905, questioned this classical perception of causality and simultaneity. (6.4)

Brain-centrism asserts that there is no way to measure causality "out there," causality is an abstract concept describing an operation of the verbal mental space. Because we represent, fragment, and abstract, we have separated an "action" from the body it is acting on. Just as we separate hardware and software. Causality is the mental operation by which the mental space "re-glues" timely what we have separated. It allows the verbal world to produce a coherent and communicable whole.

This "causal" organization of the verbal world favors memory retrieval. We associate memories with timestamps and organize them "coherently," like a puzzle. This facilitates understanding and explanation. We form causal chains by connecting past events to present situations. For any representation (R), Net (R) will include causal chains we have memorized[34]. These constructions will also rely on expectations, and fill-in to select the most appropriate causes. Therefore, different people will come up with different causal explanations for the same event.

We will also abstract the structure of a causal chain as an order or a pattern. It will allow us to use a similar procedure for different circumstances, like an algorithm.

Proper usage of causality does not seem to be imprinted in the brain's circuitry and we will need to be trained during childhood to logical thinking.

Building a convincing causal explanation is easier "a posteriori" after the event. Causal chains are not the proper tool to predict because of the variety of causes that one can select. However, by carefully learning patterns, we gain experience that can be useful for prediction.

Democritus (460—c. 370 BC) already understood that causal chains can be problematic. They are infinite and recursive. He was the first to stop the recursivity of the chain *"what is it made of"* with his idea of atoms. We remember Socrates for his doubting. What can we know? On what foundations rests human knowledge if answers are part of infinite causal chains. Ultimately, if knowledge implies knowing the causes, we end up knowing nothing.

The "origin problem" that appears with all causal chains remains today as it was in ancient Greece.

By considering causality as an operator of the verbal world, brain-centrism put forward limitations of any formal system, allowing self-referrals. (6.10)

Brain-centrism bases knowledge on the natural world. (4.2)

**

---

[34] Net(R) is the *logical context representation network*.(1.7)

3.10 The mind and body problem

René Descartes's landmark book, *Discourse on the Method of Rightly Conducting One's Reason and of Seeking Truth in the Sciences,* published in 1637 includes his famous quotation, *I think therefore I am.* It is one of the most influential philosophy books of all time. In the Discourse, Descartes, 2000 years after Socrates, starts by doubting everything. He notices that the only thing he could not doubt is that he is thinking. Thinking is primordial to any knowledge, a perfectly brain-centric idea.

After having individuated two types of "substances" mind and matter, Descartes raised an important question: how could these two substances interact. He could not answer, but we consider him the founder of modern dualism. Descartes's question had ancient sources. We can trace dualism to the Greek philosophers, especially Plato and Aristoteles. It is the belief that we must separate the world into two types of realities, material and immaterial or spiritual.

Aristotle, in his book *"De Anima,"* was one of the early thinkers asking how to relate mind and body. How to establish a connection between "immaterial" thoughts and matter?

In his Sixth Meditation, first published in 1641, René Descartes expresses the central problem posed by confronting mind and body.

"How can things so utterly different in nature as a thinking being and corporeal being, each acting and acted upon in incommensurate ways (thought vs. motion), enter so intimate union—a union that includes causal interaction between them? How can anything that acts only by thinking exert effects on something that can be acted upon only by being moved? And how can anything that acts only by moving exert effects on something that can be acted upon only through feelings (sensations, passions)?"

"I was not able to give any explanation of all this, except that nature taught me so. For there is absolutely no connection (at

least that I can understand) between the tugging sensation and the decision to take food, or between the sensation of something causing pain and the mental apprehension of distress that arises from that sensation."

Descartes believed it would never be possible to solve this problem for science:

"*In any chain of physical causes, one can only find one particle or set of particles communicating their motion to another particle or set of particles: so that even if the chain were to proceed to infinity, we would never reach anything but still other motions of particles, i.e., we would never reach desires and aversions, distress and delights, etc.*"

In 1640, the Princess Elisabeth of Bohemia, the daughter of King Frederic of Bohemia, had asked Descartes for advice in the awkward moments she was going through. She was living in exile since the 30-year war. She maintained a correspondence with Descartes until he died in 1650. The princess, in these epistolary exchanges, insisted that Descartes should offer her an answer to the mind-body problem. She could not accept the idea of two different kinds of natural entities without a clear explanation of how they could interact.

In her letter, dated from La Hague May 16, 1643, she says:

"*... begging you to tell me how the human soul can determine the body to achieve voluntary actions. Because it seems that any movement is made by the pulsing of the thing moved, in the way it is pushed by the thing moving it... That is why I am asking you for a precise definition of the soul...*"

Descartes answered on the 21st of May. He explains how objects are attracted to the center of the earth with no contact being visible between the planet and the drawn objects. He then argues that this analogy explains how mind and body could interact. The princess answers on the 20th of June in a long and detailed letter that she does not understand Descartes's analogy, and requires more explanations.

For many thinkers, Descartes's problem remains open today. Many scientists are materialists during work hours and become dualists after. Brain-centric answer on the mental causation problem is in (3.11).

Phenomenology is the study of conscious phenomena, mainly from the first-person perspective. It developed with Kant and Schopenhauer and was given its independent status as a branch of philosophy in Europe during the 20th century under the impulse of Husserl, Heidegger, Sartre, Merleau-Ponty, Jacques Derrida, and many others.

Arthur Schopenhauer was a great admirer of Immanuel Kant. Relatively unknown during his lifetime, he did not give value to social encounters and was as many brilliant thinkers, a solitary man. He wrote:

"A man can be himself only so long as he is alone; and if he does not love solitude, he will not love freedom; for it is only when he is alone that he is really free."

At the beginning of the 20th century, his work gained an enormous influence among scientists, philosophers, and artists.

He was the first major Western philosopher to take eastern philosophy seriously. Nietzsche praised him for "his honesty, his cheerfulness, and his steadfastness." Karl Popper (1902–1994) the great philosopher of sciences considered him to be: "a man of supreme integrity who cherished truth beyond anything else."

In his major book *"The World as Will and Representation,"* Schopenhauer recognizes Kant's distinction between the world as it is and the world as we see it.

"Kant's greatest merit is the distinction of the phenomenon from the thing-in-itself, based on the proof that between things and us there always stands the intellect, and that on this account they cannot be known according to what they may be in themselves."

For Schopenhauer, our cognitive apparatus, our brain, the very element that allows us to know, forbids our full comprehension of the objects it describes.

In April 1837, the Royal Norwegian Society of Sciences offered a prize for the best essay on the following topic: Can the freedom of the will be proven from self-consciousness? Arthur Schopenhauer won the prize. The problem he attacked in his essay was precisely Descartes's question: the problem of *mental causation.* (3.11) How can one reduce the mental to the physical? Schopenhauer believed that the deepest explanation is not a physicalist's explanation, but a mental one.

For Schopenhauer, the "nave" realism at the basis of the post-enlightenment sciences was a "childish mistake." He agrees that "matter precedes mind," but he also argues that in some sense "mind precedes matter." He calls this "an antinomy in our faculty of knowledge." His solution is to follow Kant's distinction:

"Time, space, and causality are forms of appearance; they belong only to phenomena, not to things-in-themselves."

Schopenhauer has this way delegitimized Descartes's problem. The price for this was high, he had to abandon ERP and materialism. He paved the way for brain-centrism.

With the development of ERP neuroscience, Descartes's problem took a very concrete aspect: How does the brain activity generate experience? What, in the brain, produces consciousness, and wherein the brain does lie? (3.11)

David Chalmers in 1995 has famously called this same mind-body problem the *"Hard Problem"* of neuroscience. He was trying to reconcile a physical phenomenon, such as brain processes and experience, phenomenal qualities or qualia generated by the mental space.

As did Descartes, Chalmers is seeking a description of a first-party inner perspective within a third party ERP framework, individuation's concepts, and language. As Descartes had already noticed, a true explanatory connection is impossible, as we have shown, even if she can observe some correlation for motor activity. The hard problem disappears if one abandons ERP who requires us to reconcile a first party and a third party perspective.

For brain-centrism,

On one side, there are the representations of the natural world described under a first-party inner perspective that we call "what we experience." They mostly operate in Gödelian information. From this perspective, we are blind to our brain activity and perfectly see what we experience.

On the other side, there are the representations of the verbal world described under a semi-third-party perspective that we call scientific observations. They operate with Shannon's information. From this perspective, we are blind to what the patient experiences, but can imprecisely measure some of his brain activity.

We should notice that these two sets of representations are obtained differently. They don't even represent the same territory. Internal Gödelian information feeds the first set of representations (1), by sensory perception, memory associations, individuations, and entailments dictated by the language; it highly depends on the state of the mental space. Limited Shannon information, collected through measuring instruments, feeds the second set of representations (2). It concerns other individuations consensually approved and following the scientific language and methods.

The tentative reconciliation is not between the tangible and the non-tangible; it is about two sets of representations that ERP wrongly hypothesizes as being two aspects of the same unique thing. For brain-centrism, we base these two sets of representations on different individuations and they result from different perception mechanisms. One cannot be explained in the language of the other. We already know that the complexity of

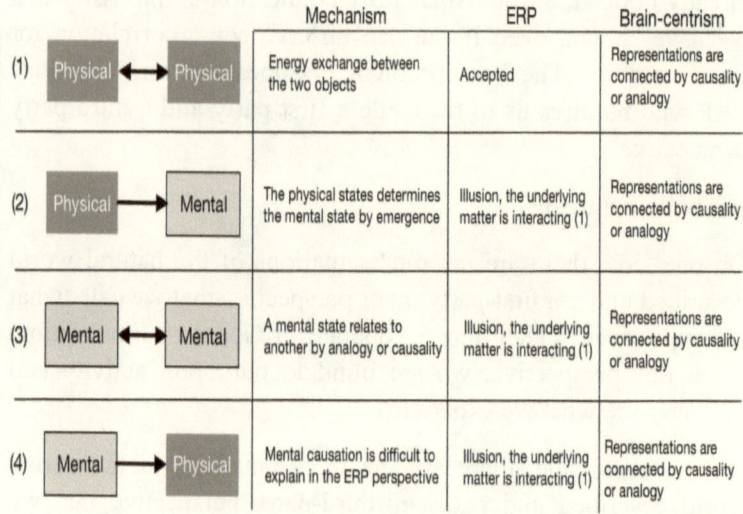

Figure 3.11.1: Mind and body problem, interraction of mental and physical

the mental space (1) is infinitely higher than the complexity of the scientific explanatory system (2).

I would like to mention the famous American philosopher and a critique of physicalism, Thomas Nagel who argues that some phenomena are not to be grasped from an objective perspective.

"We are wrong to take modern science to be the only paradigm of objectivity. The kind of understanding that science represents does not transfer to everything that people would like to understand."

Nagel is an advocate of the idea that consciousness and subjective experience cannot, at least with the contemporary understanding of physicalism, be satisfactorily explained using the current concepts of physics. In 1974 he presented an article detailing this position entitled: *"What is it Like to Be a Bat?"*

When a musician composes a masterpiece, a mathematician solves a problem that has been open for 200 years, a man overcomes a serious debilitating handicap by the strength of his will, when we travel to the moon or a child suddenly shows by his smile that he understands and loves, we must be more than a bag of molecules.

The very tools of digitalization, symbolic representations and experimentations that have allowed us to acquire so much mechanical knowledge also limits our investigation capacities when we try to apply them to study first party experiences for which their individuations and methods do not apply.

**

3.11 Mental causation

Causal efficiency is a concept referring to the ability of "something," physical or mental, to act on another mental or physical entity. How can information in a mental space be causal efficient?

In ERP science, "acting on" implies an exchange of energy between the two parties. Something physical can act on some other physical thing, like two billiard balls colliding or an electric current flowing through a copper wire. They exchanged energy. (Figure 3.11.1) (1).

A mental representation can act on other mental representations. This is the learning process we can all experiment with. This happens in the mental space by analogy or causality as we described. (3). At the brain level, this corresponds to reinforcing connections, creating new connections, and neurogenesis.

How does a physical entity act on a mental representation and a mental concept act on a physical body (2) and (4)?

Weak emergence explains (2) how the physical acts or generates the mental, although it does not explain the content of the mental picture. We have discussed weak emergence in (3.8).

The process describing (4) mental causation is straightforward. However, two things make it complicated. The first is the language that does not distinguish levels of abstraction and thus

THE VERBAL WORLD

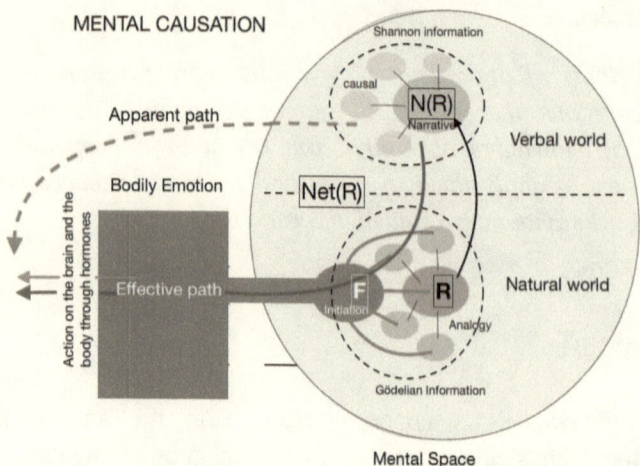

Figure 3.11.2: Mechanism by which we feel that a narrative is acting on the body

confuses maps and territories. The second is that description happens at both the level of the brain and the mental space.

Let's take it in a step-by-step description of (4), *mental causation*.

1.- A feeling (F) will initiate a representation (R) in the natural world

2.- A narrative will translate and abstract (R) in a verbal world representation N(R)

The feeling (F) is Gödelian information initiator of (R). The narrative N(R) is Shannon's information in the verbal world, symbolizing (R).

3.- The narrative N(R) is causally related to other memorized narratives in the verbal world. R is related by analogy to other natural world representations. This generates a network of called-up representations in both worlds that we will call Net(R), the "logical context representation network."

4.- Net(R) is associated with a series of feelings in the natural world whose Gödelian information will combine with F in a new feeling Net(F).

The Relativistic brain theory explains how this combination happens as analog field computing. (3.1) The system is a CAS continuous interactions occur between the analog and the digital computing systems.

5.- Net(R) is Godelian causal effective physical information related to our physical bodily emotions that act on our body. The body can then act on "out there." We interpret this whole mechanism as mental causation (4). (R) acts on "out there."

Figure (3.11.2) illustrates the mechanism we just described.

For ERP physicalism Gödelian information being integrated and thus not measurable, it is denied. ERP then asserts that mental causation is impossible and is an illusion. Therefore, most scientists are "materialists." For them, only one kind of substance can exist: matter, and one kind of information: Shannon digital symbolic information.

The procedure we described respects the laws of physics and conservation of energy. However, it extends the notion of information beyond symbolic information to integrated information. The fact that the procedure is initiated physically in feelings could be interpreted in favor of physicalism. However, conscious thinking interferes through the associations in the verbal world that make up Net(R), the *logical context representation network*.

We can thus train thinking at the level of these associations and at their meta levels. We don't need to learn every procedure, we can meta learn how to build procedures.

Intuition is illustrated by the natural world part of Net(R) that will generate analog associations. (6.6) In (3.13) we examine how Libet's experiment supports this description.

This mental construction seems to be much more coherent than denying altogether the existence of consciousness or of a mental space that we can all resent in a first-person perspective. Denying it because of the limitations of the symbolic representation seems absurd. All the more when we know the verbal system itself that has blind spots causes these limitations. (6.9) and (6.10).

Abstract thoughts or narratives cannot be causal effective according to energy conservation laws. A thought or a narrative has no energy, it is not a physical object. A story itself cannot be causal. One can write it on paper, as it is "detachable" Shannon information, but it will "do" nothing. It is only when a story is read and has triggered an emotion that it "means" something for someone. It's not the words of the poem or the notes of the music that make us feel happy or sad. The emotions to which we mentally connect them handle our bodily reactions.

**

3.12 Cartesian dualism and fragmentation

Descartes's division of the world into two types of entities is a verbal world fragmentation named dualism. Both physical and spiritual substances are mental representations. The mental representations of these two types of substances can interfere as they are mental. Descartes's problem with dualism was related to ERP. For ERP this mental interference corresponds to an interference "out there." However "out there" mental and material is not fragmented. The separation only happens in the verbal world.

Once fragmented, we have difficulties reuniting them in a common understanding.

If fragmentation has little consequences when describing non-organic isolated systems, it fails in describing mental operations, because we are fragmenting the tool that is fragmenting. (6.9) and (6.10)

THE VERBAL WORLD

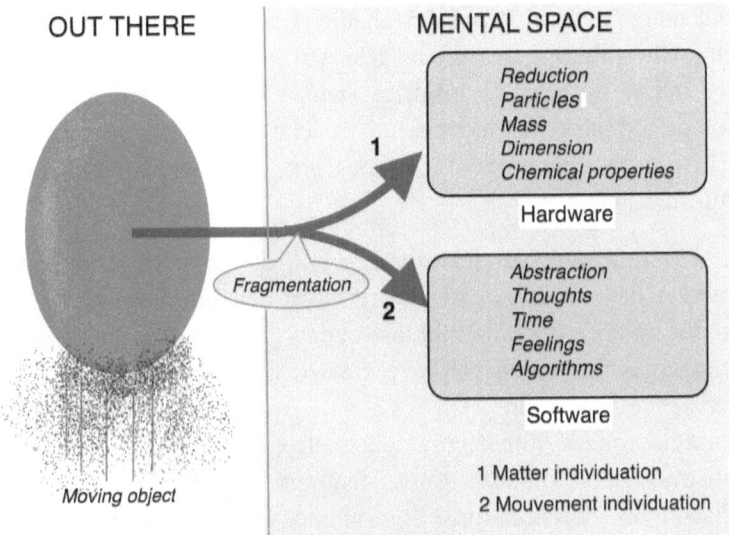

Figure 3.12.1 Fragmentation and individuation

Not that easy to understand and admit that this fragmentation/separation results from our mental space operations. There can be no "natural fragmentation out there." We fragment to represent, and we represent to understand. Fragmentation is a characteristic of the map and not of the territory. The difficulty arises because, to analyze and understand the nature and origin of fragmentation, we use the same tool that had caused the fragmentation: our structured language.

Simplified individuations induce the mind and body, "separation." When we describe any object separating it from the forces that act on it, we fragment. We have expressed that by the sentence:

Planets orbit the sun without knowing Newton's laws.

Let take our protein folding example (3.7).

We have individuated the protein as the "matter" that makes the protein its amino acids. The matter is the specific individuating quality. The folding, the action, has not got the quality of being material, it is individuated, like software separated from its material hardware. It is "created" by the mental space because out there "action" is not detachable. We then try to figure out a rule (the folding) that would take the protein (the matter) from its initial state to its ultimate state. Out there the physically integrated proteins information reaches it results in seconds. This is the essence of the concept of Gödelian integrated information.

Our fragmented individuation results from how we produce the mental map; it is not on the territory. On the map, narratives about physical and mental events use entirely different languages for which we have created mental concepts with no existence out there. On one side, you have reductionism, particles, mass, dimensions, chemistry; on the other, you have abstraction, thoughts, time, feelings, algorithms. With so dissimilar individuations, it is impossible to reconcile these narratives, although on the "out there," they are not separated. This impossibility is at the heart of Descartes's problem, Chalmer's Hard problem, or the brain code problem.

* *

3.13 Libet's experiment

The famous experiment conducted by the neurologist Benjamin Libet in the 1980s has often been used to support physicalism and the idea that consciousness is an illusion. Brain-centrism considers a key to understanding thinking and the mental space.

Libet shows that the neurological sign of a decision can be "read" in the person's brain activity some 300 milliseconds before he even becomes conscious he was deciding. The volition process thus started unconsciously. (Figure 3.13.1.)

Let's see how this supports and justifies the idea of an expectation mechanism. At the moment of the conscious perception, and before the motor act corresponding to a visible decision, one can notice a delay of 100 ms. During this Gap of 100 ms, consciousness has the possibility to interfere and change the unconscious impulse associated with the expectation. That

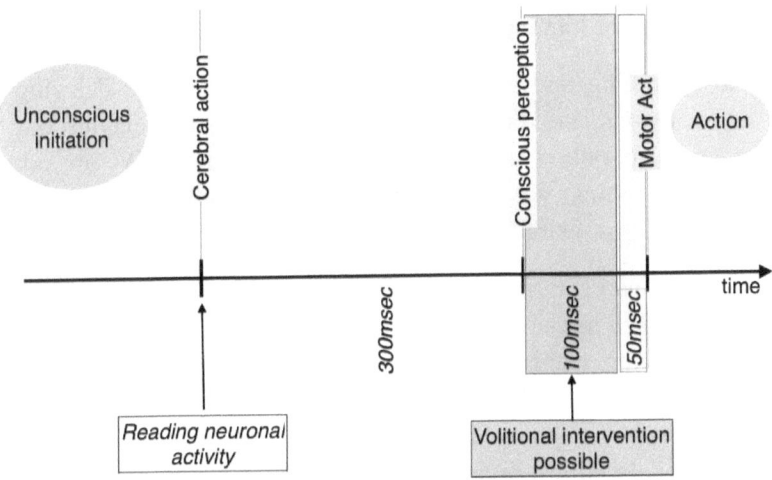

Figure 3.13.1: Libet experiment

precious 100 ms allows consciousness to delay and change the reaction proposed by our expectation mechanism. They have allowed human civilization to develop. Libet experiment thus supports our description of mental causation (3.11).

During this 100 ms, consciousness can act to accept the unconscious anticipation and the preset behavioral expectation patterns or to refute it, allowing then other anticipation to manifest and a thinking process to start. If consciousness does not take advantage of this 100 ms, the "automatic" pre-learned expectations and behaviors will be in control. It happens, in all our "automatic" behaviors, like for instance, when we drive a car or when we autocomplete a word after just reading the first letters.

Nature has left this opening to start a thinking process without imposing it. It would have been too slow to guarantee our survival. Without this 100 ms, we could never have been able of third-party perspectives.

Our shamans 30,000 years ago took advantage of this 100 ms to delay his spontaneous reactions, allowing him to explore different mental pathways through conscious thinking. This opened up for him the creation of invisible worlds.

Most learning and materialization processes involve delaying our reactions and start thinking. But for humans, delaying our reactions is only a possibility, it is not mandatory. We are not obliged to think; we have 100 ms to choose to do it; otherwise, usual patterns take full control of our reactions. (Figure 3.13.2).

Many human activities are better accomplished after a learning process that has automatized our reactions to speed them up. Like playing tennis, driving a car, or piloting a plane, doing repetitive work. In these activities, we have to learn an algorithm that will drive preset learned patterns and allow immediate reactions. In all life-saving situations, thinking is too slow, as for our primitive man in the savanna.

The final 50 ms shown in the picture (3.13.1) is the time needed to activate the muscles. During this interval, no more possibility exists for the spontaneous act to be stopped, it will go to completion.

Libet himself asserts:

The finding that the volitional process is initiated unconsciously leads to the question: Is there then any role for conscious will in the performance of a voluntary act.

The conscious will decide to allow the volitional process to go to completion, resulting in the motor act itself. Or the conscious will block or "veto" the process, so that no motor act occurs.

For brain centrism, Libet's experiment confirms that the expectation mechanism is at work throughout the mental space. Consciousness can decide because of this 100 ms delay that allowed thinking, developing third-party perspectives, and human civilizations. Educating our mental spaces to make usage of this 100 ms is critical. Even if, for economic reasons, it can sometimes seem to be a waste of time. We should let machines do algorithmic repetitive tasks, they have no way to stop their algorithmic procedures and start thinking.

3.14 Why the mental space

We have examined part of the huge abstract territory that makes up our mental space. With some of its natural and pre-existing lands, some of its older and more recent human structures, its two information sources and information systems, and some of its forces. We have emphases on the perception mechanisms that relate the mental space with the external world and allow it to update and adapt. The complexity of the mental space is such that it escapes any verbal description. A verbal description would be part of the verbal world, which itself is only a recent part of the total space. Only metaphors can give us a feel and help the mental space visualize itself.

From the outside, what we can "discover" is the brain, the physical space. A third party cannot "recognize" the mental space we experience from the inside. This is per se remarkable and has played a critical role in the way human societies have formed. That our internal world is our secret has conditioned the way we structure our society and represent our ultimate personal freedom. Amazingly, nature has developed this internal feeling of freedom by building an impenetrable wall around aspects of our mental space. Organized groups, societies, powerful men, militaries have always attacked this freedom that they consider as a potential disorganization source or a threat to their power. The neural code tentative is just but the latest assaults on our freedom wall. Previous attacks used much slower and less efficient methods, such as manipulating emotions and narratives to condition beliefs.

To describe the mental space, we could have used the language of psychology. But that language has developed as a third-party description. Psychologists dealing with suffering patients have built these individuations and concepts. Here we have other goals, we are trying to understand our reality and how the mental space generates it.

In our James Bond example (1.4), the two perspectives (the spectator and the technician) don't experience the same thing, although they are looking at the same "object." If we adopt a third-person ERP perspective to describe the mental space, the picture would be the one of a technician. It would not allow distinguishing some elements important for brain-centrism. Brain-centrism wants not only to know how the mental space "functions." It wants to understand why it creates mathematics, why it loves music, why a painter is ready to put his life at risk to finish his painting, why we build temples and start wars. It wants to understand why humans create civilizations and why they have built this reality rather than another one. Would another one be possible? What are the limits? Are there any limits? For brain-centrism, a "physical theory of everything" cannot be a theory of everything, as it would be included in the verbal world.

We thus need to include in our study the full richness of the first-person perspective, and confront it with our creation of "semi-objective," semi third party visions. We should not close our eyes to the fact that the world we live in is not an "objective" third-party world. It's the world our mental spaces build. By neglecting this, we take the risk of building an external world to which we cannot adapt. Nobody reasonable would build a building without knowing who will live in it. But we also confront the huge difficulty of reconciling two very different perspectives, the inner first-party view and the outer semi third party one. The human mental space is the only place in the universe of which we can know this double inside and semi outside view.

A photo camera is never in the pictures it takes.

The fundamental difficulty is that we need the mental space to visualize itself. When we focus on certain regions, we create recursive loops. These recursive loops generate blind spots that limit our understanding and obscure our vision. The "natural brain's reaction" when we come to such blind spots is that the brain proposes fill-ins. Observing a system from the inside creates limitations. ERP neglects these limitations when it adopts an outside third-party approach. ERP has to attribute these limitations to "out there," when brain centrism sees them in the structure of the mental space.

For instance, brain-centrism relates certain intrinsic limits of formal systems to blind spots of the verbal world. These limits can produce emerging properties or "indeterminacies." ERP, sometimes like in quantum physics, considers these indeterminacies as properties of nature out there, while brain-centrism will attribute them to mental space blind spots.

Self-referral, at the source of blind spots, makes it impossible to define primitive words by description. For example, we cannot define consciousness, existence, time, or space. To understand these words, we need the natural world and thus the first-person viewpoint that science excludes.

Many difficulties inherent to the mental space are because of self-referral or indeterminacy, here are some of these problems:

Limitations to scientific knowledge, including the measurement problem, limits to empiricism.

Reductionism and individuation.

The "origins" problem. How do things start?

The inevitability of moral and ethical questions.

Confirmation biases.

For brain-centrism, perceptions are central to how the mental space builds realities. The word perception here would include both sources of information. This remark gives a critical role to the mental mechanisms that assemble and reorganize pieces of information to build realities. Most remarkable is the role of the expectation mechanism at work at all levels between representations. Expectations are the sources of self-referral biases, they guarantee tour quest for the truth will remain endless.

Fill-ins then play a central role by filling the gaps in our first- and third-party knowledge. In the past fill-ins have determined the destiny of civilization, they have provided answers to blind spots which have oriented our hopes and our beliefs. Expectations have cut short the idea of absolute knowledge and absolute truth. But they have allowed learning at a level no other species have reached once combined with structured languages. They have also allowed incredible abuses by allowing beliefs to prevail on observations, and discourses to prevail on precise information.

The expectation mechanism has another important consequence. When no sensory input is available to change or support our expectations, these prevail and we treat them as perceptions.

It is the case when we close our eyes let our minds wander. This specificity has consequences on how our mental space treats non-observable objects. For instance, elementary particles such as electrons, photons, or quarks are non-observable. They are equivalent to their formal definition. Mathematical expectations apply then with no direct possibility for us to adapt any individuation. This has consequences for our interpretation of quantum phenomena.

Mental blind spots, because of expectations, are of critical importance in our quest for truth. It is the case when associated with fill-ins and bias confirmations. In chapter 6, we will give a closer look at them. However, we can already realize that their consequences are more important to humanity than the regions of the mental landscape more susceptible to an "isomorphic" description on which ERP concentrates.

Brain-centrism claims that studying the intrinsic limits of the mental space and its consequences should become a central research and educational task. The famous Socratic "Know thyself" touches the core of blind spots. It justifies our description of two lines of development for humans. One should know not only the object but also the characteristics and limitations of the "knower."

Nature and evolution have provided our mental space with adaptation and learning mechanisms, allowing us to overcome or circumvent certain blind spots. We can also attribute many mental states to ourselves and others. We can "guess" the intentions, hopes, expectations, desires, happiness, and beliefs of other humans and certain animals. Psychologists use the term "theory of mind" to denote this amazing ability.

Among these abilities, the most powerful one is imagination—the capacity to conceive and explore alternative realities.

As Descartes observed, we know we are conscious because we observe and know we observe. Consciousness is self-referral. We know consciousness because of what it does, not because of what causes it. This internal projector lights up the parts of the scene, but we cannot aim it at itself.

From an inner perspective, because the scene lightens up, we know there must be a projector. In the same way, we conclude by seeing photography that there must have been a camera, even if we never see it. That is how Descartes reaches the primordial conclusion: "I think; thus, I am."

A third party has no way to know the scene and deduce that the projector is there and working.

Consciousness is primitive; in the mental space, it can have no causes that explain it. Causality stops when confronting blind spots, just as it stops in the Newtonian system for space and time that must be pre-given primitives.

By a third-party analysis of the physical space, one can find correlations with the mental space. But we see no way to explain our inner experiences with electrochemical events in the brain. The language of science and its individuations and its integrated ERP perspective prevents us from complete access to the mental experience by using this form of Shannon world explanations.

Representations of external objects or events are far less subject to distortions than feelings because we confront them with outward-oriented sensory inputs. However, they are subject to various constraints related to brain physiology, language, emotions, and pre-existing expectations.

The shape of the water is often the shape of the bottle.

Immanuel Kant (1724–1804) already explained that what perception tells us about the world is how it presents itself to us, not what it is "In itself." By combining information sources, the mental space continuously changes its own "state." The intensity of our active emotions and the prevalent feelings at that moment change. So do our expectations and our dominant thoughts; it changes the active mental patterns; it changes memories and the complete reactions of the rest of our body. A third-party perspective, when analyzing the brain or a stone, only feeds at sensory information.

There is an uncountable number of possible mental states, and we create new ones through brain plasticity and adaptations at every level. A mental space cannot be twice in the same state during one's lifetime. We cannot individuate a mental space independently of the environment, be it physical or social. Miguel Nicolelis has studied in his lab what he calls brainnets. These synchronizations between primates' brains occur through speaking or through listening or observing a common scene. He showed that one can establish a brainnet by sharing brain signals from "brain to brain."

At church, in the laboratory or a court of justice, at a ball, on the beach, or at a football match, each situation triggers the opening of a pattern. We do not use the same concepts and do not develop the same feelings in every life situation. The language is different, and, most of all, the balance of the various information sources is different.

How does the mental space solve all the problems we encounter, from playing chess, crossing a busy road, preparing our food, speaking, writing, building a spacecraft, greeting a friend, choosing how to answer a question?

Jerry Alan Fodor (1935–2017) was an influential American cognitive scientist and philosopher. For him, "modules" structure the mind. Each has a specific function and a geographical location in the brain. Every module has specialized problem-solving skills, a module for language, another one for perception, another one to control our emotions, a module to recognize faces. A central processing part takes care of the relations between the various contents.

The most compelling evidence for the modularity approach comes from studies of patients with damaged brains. Damage often impairs some cognitive capacities, but not others. A patient can, for instance, lose the capacity to understand speech but still be able to speak or lose long-term memory without the short-term memory being affected.

But the mental space also has general-purpose problem-solving capacities. Its incredible plasticity allows the "take over" of damaged regions by other parts of the brain. Recent research, for instance, has shown that one can develop a visual image through touch and kinesthetic movement, called haptic feedback; we are all familiar with inserting a key in a lock in the dark by touch and pressure feedback sensations.

To finish this overview, I would like to remind us how complex the mental space is. Its complexity is infinitely higher than the complexity of certain of its subsystems, like language or theoretical physics. We will obtain simple proof of this assertion by noticing that the complexity of mathematics is infinitely higher than the complexity of any language, formal ones, and natural ones. This assertion justifies our metaphorical descriptions. It is then not astonishing that the underlying brain does not fit into traditional physical descriptions.

* *

Properties, functions, and components of the mental space reviewed.

- Perception, Representations, feelings, expectations, representation networks,
- Anticipation or expectation mechanism, fill-in mechanism, Initiation of expectation
- Fragmentation, Individuation, Identification, association
- Adaptation, Analogy, Causality, Order, Curiosity
- Russian dolls, adaptation levels, complexity, complex adaptive systems
- Abstraction, abstraction levels, about function, mental abstractions
- Memory, short-term memory, mental patterns
- Learning, imitation, neurogenesis
- Feelings as representations of Emotions
- Two information sources, analog and digital
- Survival, primitive emotions, and feelings
- Symbolic structured languages
- Inclusion principle
- Irreversibility
- Shannon information, digital and analog combination. Gödelian information.
- Reductionism, holism, decomposition, fragmentation, wholeness
- Emergence
- Self-reference, recursively, recursive loops, blind spots

* * *

IV: KNOWLEDGES, BRAINS AND COMPUTERS

Knowledge makes up the substance of the mental space. In this chapter, we will review how knowledge is gained. We will distinguish two types of knowledge corresponding to the two worlds we have described.

The verbal world alone cannot find out the truth of a narrative. We call this fundamental question, "The Truth Problem." We will describe the ancestral philosophical struggle to solve the truth problem. This millenary story is critical to understanding reality.

4.1 Studying with Pythagoras

It was not so easy to enter the school of Pythagoras settled in Croton, in southern Italy, about 525 BC. An applicant would be refused and put under observation. That observation period would last several years. If they admitted him, it was for a five-year vow during which they would teach him music, geometry, and magical incantations. Pythagoras emphasized moderation, piety, respect for elders, and of the state. He advocated a monogamous family structure and preached an austere lifestyle. Pythagoras believed in social justice; justice had for him its origin in geometry. He also believed in the reincarnation of the soul and the afterlife. Humans could reach reincarnation through their conduct during their life on earth. Their soul was then admitted to another world. Aristotle explains that Pythagoras, on the tracks of his teacher Anaximander, could calculate the size of all the known planets and that, for him, all the planets were spheres orbiting the sun.

The observation period, preceding the admission, allowed the school to detect the sincerity and the deeper motivations of the future student. No admission exams were required, and they delivered no diplomas. Most of what the student would learn, he would learn it by an acquaintance, and not by verbal discourses.

Pythagorean ideas would exercise a marked influence on Plato and Aristotle and through them, on all Western philosophies. His school had become famous for some of its basic ideas, including

The nature of reality: Reality is, at its deepest level, mathematical. Music and astronomy are examples of how numbers underlie everything in our world. Numbers were more than quantities; they had more profound importance as guides to the harmony of the soul.

How to make one's mind clear: they considered philosophy a guide to spiritual purification. Without a purified mind, we lose track of our internal guidance. This purification needs to be learned; it comes by itself. Meditation did the learning; the student had to discover by himself before any theoretical discussion was possible.

The human soul can unite with the divine: The human soul can reach unity with what transcends us, but it does not come automatically. Purification is the way to rise to this union. Uniting with the divine means a deep personal global understanding, leading to personal peace and wisdom in interpreting nature and society. It was the ultimate goal of the human mind. It was considered that, without this union, any understanding would be limited; any conclusion could further down the road lead to disasters. Narratives were not the method to discover the divine, it only "appeared" after sufficient guided meditation on symbols. The divine unified all knowledge and defragmented the student's personality.

The study of spiritual symbols: Students would learn the symbolic nature and meditate on the beauty of numbers and relations such as the golden section and the harmony of the spheres. Although the matter had technical roots, this study was contemplative. Each symbol, each number, could help develop an aspect of the soul.

The school promised nothing; it gave no diplomas or certificates; it did not teach how to "make money" or become famous. It did not consider the purpose of knowledge as related to power or social status. It warned on the fact that power or the quest for honor would corrupt accurate knowledge. The philosophical program of the school aimed at liberating from the animal's passions and identifying our divine part.

However, the central thesis of Pythagoras of Samos was that the number is the principle and the origin of everything.

Reality is much too complex to be expressed by one unique principle, Pythagoras separated reality in two opposed principles: the finite and the infinite. Numbers are the key to understand how these two principles interplay and combine to generate reality. The odd numbers represented the infinite principle and even the finite. Equilibrium resulted from the proper balance between these two principles; without this equilibrium, there was no hope to concentrate on purifying the mind and the soul.

Interesting is that for Pythagoras, numbers represented not only quantities, each had a symbolic meaning and a moral signification. It had to be discovered. By contemplating numbers, their combinations with their symbolic meanings, the student would go through his purification process.

By referring to numbers from two perspectives, the school of Pythagoras elevated its students for them to find internal guidance; by associating practical techniques with spiritual meditation. By taking the time to reach a deeper understanding of the "hidden orders," were among Pythagoras methods to emphasize equilibrium between inner and outer sources of information to purify the minds. To understand the first-party perspective, one had first to develop the third party perspective.

What we consider today the legacy of Pythagoras; his theorem on the right triangles was for him a minor technical aspect. At his times, his most celebrated idea was the discovery of the divine that unified the personality by equilibrating inner and outer knowledge.

Pythagoras teaching and methods might seem strange to us today. They reflect a culture and values that have largely disappeared. What we have kept from his teaching fit in our present culture.

* *

4.2 Knowing and Being

> *Education is not the learning of facts, but the training of the mind to think...* Albert Einstein

So many things have changed in our societies since Pythagoras times. It can be because of environmental changes, like earthquakes, volcano eruptions, falling meteorites, biological mutations. Many are because of deliberate human judgments, like wars, alliances, regulations. Others are because of possibilities opened up by technological innovations, like motors, telescopes, airplanes, electricity. However, many evolutions are caused by emerging properties of these first three causes. In these cases, nobody or no group has decided anything, no invention is the source, and nobody could have predicted the changes. Cultural changes belong to this last category.

Explaining how societies evolve is delicate. Certain ideas can be successful, although they are harmful to survival. It seems like Darwin's mechanism does not account for emerging properties.

It might now seem ridiculous to meditate on numbers; we prefer studying how to use them. Materialism and "efficiency" have won a firm grip on the mental spaces. We are not living in a mystical era. We focus on the practical and technological side. It is much more efficient for engineers who have to build something. Nobody would have the patience to wait years to be admitted to the school of Pythagoras.

We have given a foremost priority to the measurable; power is now the major reason we want to gain knowledge. We teach our students how to perform what society will demand from them.

Knowledge has become "how to do," rather than "how to be." "How to execute," rather than "how to question."

What made us take this direction, is it a determined direction that any civilization would take, the brains have reached the intelligence plateau? Is it a necessity imposed by Darwinian Competition?

Western education systems favor *descriptive*[35] *knowledge* methods. The knowledge that description does not cover has to be gained as an inner first-person perspective. This knowledge concerns our natural world. Feeding the verbal world with information does not help. Let us cite, for instance, the knowledge one gets by practicing meditation or yoga, the knowledge taught by Pythagoras, where his students gained wisdom by meditating on numbers. But also the evaluation of moral and ethical issues, the fairness issues, truth and trust issues... Knowledge like those developed by our shaman painter who had learned how to master his fears and balance his emotions. The knowledge one may gain by spending a night outside looking at the stars. Knowledge like the one's interactions with an animal can give us, or like spending time in a different human culture. Knowledge gained by discovering alone proofs of mathematical theorems. Knowledge of our limits. Knowledge on meaning, on feelings, on love.

The enlightenment in the western world has brought forward the preeminence of reason and indirectly an ERP worldview. This has resulted in the science and technological revolutions that we have known. In Asia, Zen has focused more on the development of perception. This has resulted in higher degrees of inner knowledge or understanding that we could call "wisdom."

We collect "out there" 20 to 30 musical notes, our mental space, our natural world, then generate the whole opera. Our life is the opera, but we could not live it without the notes. It does not learn the notes; we have to study how, with those notes, we can generate the most delightful opera to live in.

---

[35] See (3.6) for learning by description or by acquaintance

Collecting other notes or having a different mental space organization would generate another opera in which to live. Between two humans, the operas differ. It differs even more between us and another animal or an extraterrestrial. However, between humans, we can agree on the notes; repeated experimentations can define and gauge them. But they do not assure fulfilled and balanced lives. Considering that our reality is only the notes are not sufficient. If we could only learn notes, only data, no meat around the bones, no meaning even for the bones, we would be computers.

Informational data are like pebbles falling. Only studying the pebbles is not sufficient, because falling on various surfaces will produce different reactions. The ripples landing pebble produce will not be the same on every kind of surface. The ripples are not the stone. Mental spaces produce an immense variety of ripples.

Their teaching is easier to organize, it allows to control and measure the results, and doesn't need to be adapted to each student. One can teach facts and avoid teaching how to understand these facts. It allows the teacher to concentrate on a third-person perspective.

George Ivanovich Gurdjieff (1877–1949) was a Russian mystic and philosopher, a strange personality indeed. His teaching was about "awakening our consciousness and unifying the personality"; teaching by some aspects resembling the one of Pythagoras. It took its roots in various oriental traditions that Gurdjieff adapted to the western world. He grouped the teaching of human development into two learning categories: "Being" on one side and "Knowing" on the other. Being centers on "inner information," feelings, and acquaintance, it's learning for the natural world. Knowing centers on descriptive knowledge, it's learning for the verbal world. Only a proper balance of these two aspects of learning can lead to a proper balance of the personality and liberate our higher mental potential. (Figure 4.3.1.)

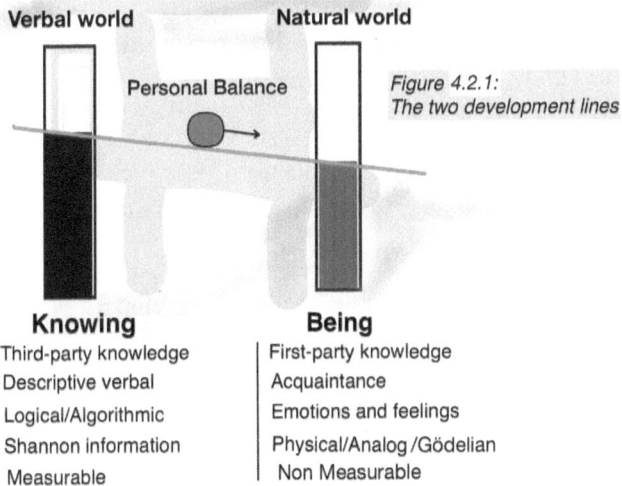

Figure 4.2.1: The two development lines

Higher are your responsibilities, more powerful is your voice, and more this balance becomes important for you and your environment.

According to Gurdjieff, one can learn how to "be more" as well as how to "know more." Both lines should grow in parallel. Knowing more means accumulating factual or technical information. Gurdjieff claimed we could also act on our unconscious patterns, and we have to do so. If emotions are difficult to change consciously, we can change the feelings that represent them.

Both development lines have to grow in parallel to gain, "actual knowledge." (Figure 4.2.1.)

Accumulating "knowing" without having developed "being" creates a personal imbalance. One could lose the sense of purpose or run after unworthy goals. Here, one is acting hypnotized, in a robotic way. "Being" without the "knowing," leads to inefficiency and incapacity to act on the world. It generates a lot of frustration.

The knowing is always a third-person perspective verbal world knowledge. The being is a first-person perspective, natural world knowledge.

The natural world expresses mastery of one's fears and negative feelings, our self-image, the sense of integrity, dignity, of truth, our various human values, our mental attitudes on the "being" line, using Gödelian information.

We left the family alone in charge of developing the "being" of our children. Very knowledgeable people can have very little wisdom. Their knowledge and diplomas put them at the summit of social hierarchies, where their lack of being qualities can cause significant damage to society.

Cultures differ one from the other only on the being line. The knowing one ends up being similar everywhere.

Gurdjieff insisted we are at a point where humans need guidance to develop the "being," just as they need guidance in the development of the "knowing." The challenge of humans and humanity is not only to master nature "out there," but also to master themselves "in there." Because what happens in there determines the interpretation and the usage of what happens out there.

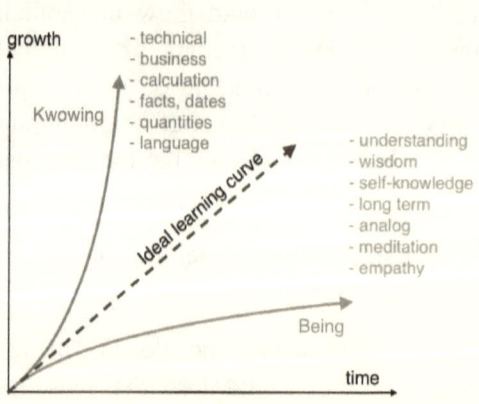

Figure 4.2.2: The ideal learning curve

In a society dominated by descriptive knowledge, knowing develops at the expense of being. The social price to pay for this imbalance starts with stress, lack of a sense of purpose, destruction of human values, mechanization of humans, lack of social fairness, social unrest, violence, lack of appreciation of our position in nature, mixed-up priorities and generalized health problems. If we already see these consequences generalize today, we are just the beginning of a catastrophic scenario. We are running the productivity loop, squeezing to increase productivity everywhere, with little respect for human nature.

Without the guidance of wisdom, the power of technology centralizes information. It destroys diversity who is a critical survival asset. We are running an insane race. Wealth concentrates in a few hands. Technology and wealth take their decisions for their reasons.

We can see this happening under our eyes now.

If an oracle or an extraterrestrial answered any question, we might have, that would in fine not be so helpful. An answer is only on the knowing line. We need in parallel to discover and understand the being one. Brain-centrism asserts that only our answers matter because we live in our universe, what we need is understanding. Only by understanding (a being asset) can we progress and ask further questions?

Changing the world is about changing the map, not the territory. We have put on the map artificial paradises, strange and destructive attractors. Changing the map will take care of us because that is where we live!

* *

4.3 Knowledge and the quest for truth

These last 15,000 years, the main vector of information, that had been sensory perceptions, has become the language. Information we are receiving and memorizing now is verbal and is feeding the knowing line. However, most of it gets forgotten.

Decay theory explains retrieval failure. Our memories fade because we have not assessed them for some time. We forget 56% of information within an hour, 66% after a day, and 75% after six days. We know we reinforce a brain connection by using it, and that we create a memory trace every time we memorize additional information. Savants prove that the brain's capacity for memorizing is huge. Therefore, if we forget, there must be an evolutionary advantage. Psychologists believe the brain prunes memories that become unused, a process that is known as active forgetting. An old unused memory would "leave its place" to new information.

What remains after some time are feelings and impressions from the natural world that we have attached to the verbal input.

Pythagoras wanted to make sure that his students were motivated, he did not want them to register for the wrong reasons. He understood he should not limit the mental space to the verbal world. The purpose of Pythagoras's meditation was to prepare the mental space to reorganize itself in order to register new concepts and sensory data.

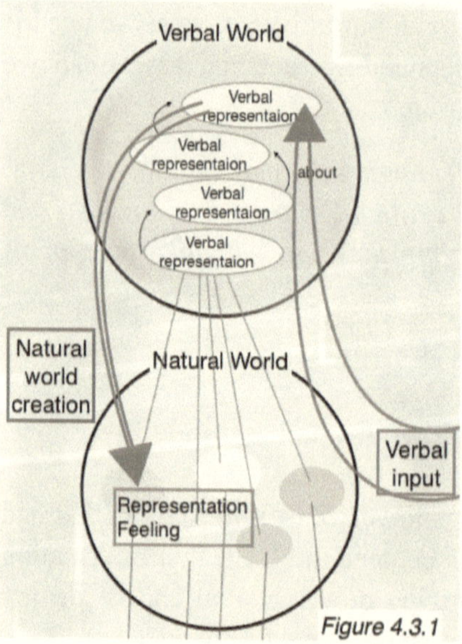

Meditation allowed his students to discover themselves the deep significance their mental space could attach to any verbal information. Pythagoras understood one knows what one has created. Learning is creating (from the first party perspective). Mental space's own creations remain for ever, memorized descriptions vanish, leaving only a vague impression.

Figure 4.3.1

Learning should thus involve the natural world. Remember the figure (2.5.1) describing the abstraction process. In descriptive learning, we associate directly the sensory verbal input to an abstract concept. To be merged, this representation in the verbal world must find deeper roots. It does that by a sort of inverted procedure. The verbal representation triggers "creative" activity in the natural world illustrated with the descending arrow on the figure (4.3.1). We form new representation networks, including feelings.

Language favored third-party views, even if the individual perception is different, the words and the narratives can be common. Therefore, teaching now concentrates on verbal world knowledge and the "being" line is left for each of us to handle.

When humans could share narratives, cultures and civilizations developed. "Reality" checks, for truth, or survival could not apply to many of the founding myths. Because the narrated events were in the past or a remote location, no confrontation with "out there" could happen. Narratives were transmitted and "believed" with no implications.

Many of these stories helped fill in knowledge gaps and create a common social agreement. Natural selection could not directly operate on these "gap-filling" beliefs; their truth or falseness did not make any immediate selective differences. If a tribe believed that the sun was orbiting the earth or the contrary, changed nothing to day-to-day life. Natural selection would not favor or condemn a population because it believed in twelve gods instead of only the ten true ones. Natural selection was not, at least in the short term, capable to act on mental abstractions.

The eternal struggle for dominance and power learned how to use language for its purpose. Rivalries among groups concentrated on rivalries on narratives. The winner of the fight would impose what narrative to believe! And the master of beliefs would gain dominance and power.

The "truth question" remained opened: Is there a way in the verbal world to determine the truth of a proposal.

An astonishing characteristic of the mental space that Aristotles would later formalize as the "excluded middle principle," played its role in selecting beliefs. We don't accept something to be there, and to not be there at the same time; our mental space does not handle contradiction, we feel uncomfortable with it. If two close enough representations contradict, we choose one and erase the representation that fits the less in our global perspective.

Reacting to contradiction has developed because of learning. Expectations have to learn from incoming data to adapt to the environment. Without the excluded middle observation and experimentation would be useless. When languages appeared, this same model of adaptation to new incoming information applied to narratives. By learning, the excluded middle has integrated the way our mental space builds our logical context representation network, Net(R).

Aristotle, in his "Organon," was the first to propose "rules of formal logic," (6.2) acting as guiding "orders," to screen and evaluate a narrative for consistency of its contents. Aristotle was trying to criticize the reasoning and the rhetoric of the sophists. These knew how to exploit language and words to entertain or persuade an audience. They understood how to develop beliefs by appealing to the public's emotions and desires. They could handle promises, predictions, secret fears, and knew how to appear as saviors. Sophists with their narratives and public speeches were putting in peril the Athenian democracy with their lies. Their rhetoric was unconstrained by any commitment to facts or truth, its purpose was to convince by entertaining and manipulating emotions.

Rhetoric allowed to gain political power and economic wealth, independent of the truth of the narrative. Teachers of rhetoric focused on controlling the audience's emotions rather than resorting to their logical thinking. Pericles was elected. He knew how to manipulate fear and comfort with his speeches. By exploiting language, Pericles became a despot and ended up destroying the first-ever democracy. Misleading speeches and manipulation left Athens a legacy of instability, bloodshed, and genocidal warfare.[36]

Aristotle's dream was to establish rules that would guarantee the proper functioning of reason. In the third part of the Organon, Aristotles develop the art of demonstration, based on syllogism that he opposes induction[37]. The primary purpose of his "logic" was to avoid the contradictions created by manipulating narratives, to "clean up" human knowledge, and fight the sophists!

The quest for truth had started much before Aristotle, but he was perhaps the first human to formalize it in the language's structure. He realized that if a sword can kill a human body, an untruth could end up destroying civilizations.

Language and the marvelous development of the verbal world showed up with a bomb in his suitcases. A bomb that Aristoteles had hoped to defuse from inside the verbal world. But although he put quite some order, he failed. He could not answer the truth question,

In a little essay introducing brain-centrism, we developed the question: *What happens with truth if everything happens in the mental space*[38].

**

4.4 Leibniz dream

Big and profound questions keep haunting us through millennia. They build through space and time a network of related mental spaces, dedicating their lives to solve them. A thinker dies, but his problems survive for another thinker to pick up, centuries

---

[36] Unfortunately Pericles was not the last democratically elected Governor who used the sophists teaching!

[37] induction, is the mental process that moves from particulars to their generalizations. It is the basis of knowledge. It implies primitive indemonstrable proposals.

[38] Que devient la vérité si tout se passe dans le cerveau, 2021 Editions Sarina

later, under different latitudes and with different perspectives. The "truth question" that structured language brought in its baggage is the most crucial one. What is truth? Is there a way in the verbal world to determine the truth of a proposal? How can we make sure that a narrative is true? Answers to any other question all depend on the truth question.

The question of the truth of narratives must be as old as language itself. But we will fix its origin with Aristotle and his fight with the Sophists about 2500 years ago. The rules of logic he established gave him the reputation of being the greatest logician of all times.

Our story here will be about structuring the language to convey truth: answering the truth question within the verbal world. We will start with Aristotles and will end with Kurt Gödel, the second-biggest logician of all times. The dream to master the truth of narratives through structural rules went through multiple different routes all around the globe, and was picked up by many brilliant minds. All, like Aristotle, were revolted minds who rejected the devastating damages lies are causing, generation after generation.

Truth, we know, does not come naturally to the human mind. We have to invent methods to compensate for our perception weaknesses. What can we rely on if our mental space, with its expectations and biases, is the ultimate judge of what is true? Are we prisoners of ever-ending loops of unjustified beliefs, where we see what we want to see?

Gottfried William Leibniz (1646–1716), though he had found a method to guarantee the truth of a narrative and solve the truth problem. Leibniz is best known today for his differential and integral calculus, which he invented in parallel, but independently from Newton. The variety of his contributions in philosophy, mathematics, linguistics, epistemology, mechanics is unique. Leibniz should be considered as the father of so many lines of thought and could be qualified as a founder of brain-centrism:

"Moreover, matter and motion are not substances or things as much as they are the phenomena of the perceiver, the reality of which depends on the harmony of the perceiver."

Two thousand years after Aristotle, Leibniz had picked up the same dream. It obsessed him during his entire life. He wanted to find a linguistic system that would ensure the infallibility of reasoning, making the correct use of language a guide to truth.

In his *"De Arte combinatoria"* in 1666, an extended version of his doctoral thesis, he explains his dream. The *"Characteristica universalis"* would be a formal language integrated within the framework of a universal logical "calculation" that he named "calculus ratiocinator." Computation capacity is then integrated into the background of language, it would then be possible to calculate truth.

The 17th century in Europe was a period of widespread conflict and instability. Some historians have called it the "general crisis." Events such as the English Civil War, the Fronde in France, the Thirty Years' War in the Holy Roman Empire, and revolts against the Spanish Crown in Portugal, Naples, and Catalonia were manifestations of this crisis. In that context, Leibniz imagined that a language generating truth by calculation could erase misunderstandings and avoid the wars and conflicts that plagued humanity.

Leibniz realized very young he had an enormous advantage compared to Aristotle. In 1140 Robert of Chester, a Spanish mathematician, had translated to Latin a book of Al-Khwarizmi, written in 830: *"Liber algebrae et almucabala."* Bagdad and the Arabs enjoyed the most advanced and flourishing culture on the planet. If the Romans had adopted the Greek Gods, the Arabs had taken over most of the ancient Greek knowledge in mathematics and made significant developments themselves.

The translation introduced Algebra to Spain and Europe for the first time. Al-Khwarizmi's book became over the centuries the standard mathematical text for Algebra at European universities. Algebra, along with the Arabic numerals, laid the foundations for Europe's Renaissance and the Scientific Revolution to come through Copernicus, Galileo and Newton.

Leibniz realized that algebra and geometry could give him a precise idea of what was for him a model of correct reasoning. Mathematics after all didn't have a truth problem. The same way Greeks had formalized Geometry, and Arabs had structured algebra, Leibniz wanted to structure language for any narrative to become provable with a calculation. He had to find a manner to formulate language in such a way that he could prove a sentence true with a calculation. And algebra was a perfect model for how he should proceed. Language had to be mathematized.

Here is how he did it. He first noticed that he could build categories for concepts, which could then organize these categories to form propositions. Once he had applied this to build propositions, he should then categorize them to form truths. Leibniz believed that by using this mathematized language, he could transform an argumentation into an equivalent to the "proof of a theorem." With misunderstandings or conflicts, one could then sit down and calculate. The language would become an "Ariadne's thread" of thoughts that one only needs to follow, step-by-step, to be guided towards correct reasoning.

If this dream seems crazy to you, it is because it's crazy. Humans can dream "a priori" unrealistic dreams, that is how we advance. Human heroes are not those who solve a problem; they are rather those who refuse a situation.

Leibniz's dream hides in our quest for guidance towards "accurate knowledge." But it also hides the hypothesis of the existence of "objective" truths out there, truths that a system of rules like the ones of algebra could establish.

To realize his dream, Leibniz pursued two considerable projects. Both projects have repercussions up nowadays.

The first project was to build a new language in which any ambiguity would be impossible. This is how Leibniz did it: In this language, every sign corresponds to an idea, a bit like in Chinese. Simple concepts would have a role similar to prime numbers in arithmetic, their combination would then allow access to more profound ideas. To construct this language, Leibniz invented the binary digits 0 and 1's as elementary components. This idea set the stage for what would become 300 years later, Shannon's information theory. He showed how, by using such a constructed language, everything could be expressed using his basic binary digits. In the 1670s, Leibniz was already laying the seeds of our computer languages.

His second project leads him still a step closer to his goal: thinking by calculating. For that, Leibniz had invented in 1672 an amazing digital calculating machine: *The Step Reckoner*. He built two prototypes; one still survives and is deposited in the National Library of Lower Saxony in Germany.

He remained convinced until his death that a symbolic scripture and proper calculation rules inspired by algebra would lead to a world of peace, where nobody could propagate thinking mistakes.

Leibniz's dream of peace never came true, he did not solve the truth problem. However, he inspired many followers, such as the mathematician George Boole (1815–1864). Boole is best known for this book: "The Laws of Thought," published in 1854. Another follower was Charles Babbage (1791–1871), the mathematician who originated the concept of a digital programmable computer with his programmer Ada Lovelace.

Babbage built, in 1822, the first digital computer, a mechanical device that he named the *difference engine*. A programmable mechanical digital computer: the Analytical engine then followed. Ada Lovelace became the first-ever programmer. (We named the Ada programming language in her honor.)

Dreams are so important and characteristic of a human's mental space. Because we can imagine even absurd things, it gives us the energy to advance, to adjust our thinking, and, perhaps over generations, discover something different from what had been the initial dream.

Dreams happen because we can close our eyes and let our mental space build on our expectations with no censorship of sensory inputs. We can be voluntarily irrational, we can become creators within our mental space. Geniuses unconsciously feel that beyond the absurdity of their dream, there is something that needs to hatch. A belief pushes them, not rational thinking.

By planning too precisely what you want to discover, you end up doing "more of the same thing," you discover nothing. The rational always start in the irrational. Creativity starts in the natural world. Imposing rationality at the beginning of a creative process leads nowhere. A deduction is always a loss of information.

Nature has provided us with a hybrid digital-analog brain and a mental space with two worlds to escape the rigid prison of rationality. Creation starts with the irrational feeling that some situation is unacceptable or revolting. It starts with analogies in our natural world. New knowledge is not a matter of rational thinking; it is rather a matter of not accepting.

Questions, hopes, and dreams are more important than answers because they keep us going; they keep us creative and alive. We need good questions; we need to find correct formulations, to ask. Good questions result from broader overviews, from larger individuations, from unexpected analogies, from unplanned events. They don't all originate through increased fragmentation and specialization or rational deduction. Today they trained us for answers more than for questions. Answers are only a pretext to find the next good questions!

From Leibniz to Gödel, Hilbert, Turing, and Von Neumann, it took another 300 years for us to mature Leibniz's ideas and to answer the truth problem. This led us to the digital computer as a side product.

4.5 A deep mathematical crisis

In the legacy of what Aristotle and Leibniz had originated, a deep crisis appeared nearly 300 years later. The notion of mathematical truth was the foundational inspiration of the structure Leibniz wanted to introduce in language. But this structural rigor, who was meant to avoid any contradiction, and to give truth objectivity, started showing puzzling signs of weakness at the very roots of mathematics. Here is what happened:

Gottlob Frege (1848–1925), a very famous German mathematician professor at Jena University, known as the father of analytic philosophy was finishing the second volume of his monumental work, entitled: *"The Foundations of Arithmetic."* Frege was basing his construction of Arithmetics on a logic extending the one of Aristotle's, and inspired by Leibniz's dream concerning provability of human knowledge. His book was weeks from going to press when Frege received a letter on June 16, 1902, from a young Englishman named Bertrand Russell.

Russell pointed out a paradox in Frege's construction layered out in volume one. Russell showed that the axioms Frege was using to formalize his logic were inconsistent, an absolute catastrophe for any mathematical construction. It is the equivalent of an apple falling upwards for Newton's laws of gravity, even worse. Frege replied on the 22nd of June, recognizing that Russell's remarks were "a catastrophe." He added at the last minute the following appendix to his second volume:

Hardly anything more unwelcome can befall a scientific writer than that one of the foundations of his edifice be shaken after the work is finished. This is the position into which I was put by a letter from Mr. Bertrand Russell as the printing of this volume was nearing completion.

Russell's letter was the first warning shot announcing the deep crisis that was about to shake the mathematical world for the next 30 years. Mathematics is based on consistency, and it cannot tolerate contradictions. One contradiction somewhere and the whole castle crumbles. And Russell had just spotted one at the very roots of mathematics: arithmetic.

Here is the paradox as Russell formulated it:

Let R be the set of all sets that are not members of themselves. If R is not a member of itself, then its definition dictates that it must contain itself, and if it contains itself, then it contradicts its own definition as the set of all sets that are not members of themselves.

We consider R in this sentence both as a set and as a member of the set. This construction is one root that builds the paradox. The same symbol, R, designates two abstraction levels, that are then mixed up confusing map and territory.

The Greeks well knew this type of paradox, the model being the ancient Cretan paradox posed by Epimenides:

This statement is untrue becomes paradoxical when you apply the statement to itself. You then create a situation where the

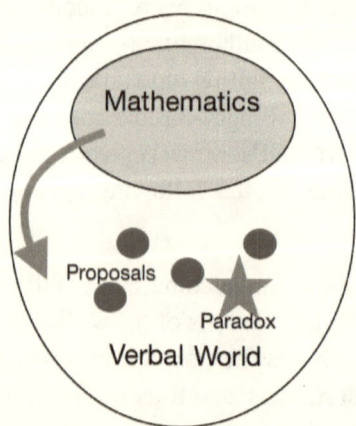

Arrow: Leibniz wanted to apply the methods of mathematics to the whole verbal world believing that all mathematical proposal could be proven true or false.

Figure 4.5.1: Paradoxes are generated by the verbal world and icontredict Aristoteles excluded middle

sentence is a statement and the object of the statement and you mix up two abstraction levels. This identification of map and territory creates a blind spot.

Another expression of this mental construct is the famous liar's paradox

The liar declares: I am a liar. If he is a liar then his declaration "I am a liar" is false and he tells the truth. But if he tells the truth, that means that he is a liar.

Paradoxes contradict Aristoteles' excluded middle principle. They are "unacceptable" for the verbal world. When a proposal is a paradox, it can be true, and non-true at the same time. We cannot answer the truth question for them. In mathematics, paradoxes are an absolute catastrophe compromising the complete structure.

* *

4.6 Gödel and Hilbert

At the beginning of the 20th century, to circumvent the crises Russell had triggered, David Hilbert (1862–1943), one of the greatest mathematicians of all times, proposed to ground all theories to a finite, complete set of axioms, and provide proofs of the consistency of that these axioms.

Any proposal written in a formal system and respecting the rules would then be true within that formal system. That was Leibniz's idea, but restricted to mathematics.

To implement his ideas, Hilbert proposed a vast program of formalization of all mathematics. It comprised rewriting all mathematics in a formal language to be sure to avoid any shortcuts because of interpretation.

The famous Bourbaki books that many students have worked on, was an indirect consequence of Hilbert's program. The program was well underway when something as surprising as Russell's letter to Frege happened to the great Hilbert.

In 1930, a young Austrian mathematician, Kurt Gödel, would prove two results that nobody had expected, and that made Hilbert's program impossible. Specifically, the second theorem.

The crisis started by Russell had shaken the mathematical world, and for 30 years, the best minds had been at work to build a solution to rescue mathematics from inconsistencies. Russell and Whitehead had produced the "Principia Mathematica," a three-volume work on the foundations of mathematics, and showed that the set of all sets could itself not be a set. Hilbert had involved hundreds of mathematicians in his vast formalization program. Kurt Gödel, working on his own in Vienna, had put an end to it by proving his two incompleteness theorems. We can express them in the following way:

1. *For any consistent formal system F in which we can carry out a certain amount of elementary arithmetic is incomplete. That means that in F there are statements, which can neither be proven nor disproven in F.*

2. *For any consistent formal system F in which we can carry out a certain amount of elementary arithmetic, we cannot prove the consistency of F in F itself.*

Hilbert was so shocked that he seemed never to have accepted Gödel's proofs. He never wrote about Gödel, never met him, never spoke about him.

This is what the philosopher Wilfried Sieg wrote about Hilbert's reaction:

In fact, there are no communications between Hilbert and Gödel and they never met. Perhaps the second incompleteness theorem on the unprovability of consistency of a system took Hilbert by surprise. We don't know exactly what he made of it, but we can appreciate that it might have been disturbing, for he had invested many thought and emotion in his finitary consistency program which became problematic as a result.

Hilbert was convinced that we could solve every mathematical problem. He had once more declared this belief at the opening of the Society of German Scientists and Physicists who held a meeting in Konigsberg, in September 1930:

"For the mathematician there is no Ignorabimus, and, in my opinion, not at all for natural science either... The true reason [no one] has succeeded in finding an unsolvable problem is, in my opinion, that there is no unsolvable problem. In contrast to the foolish Ignorabimus, our credo avers: We must know, we shall know."

Hilbert made this declaration the day before Gödel announced, at a different meeting in another part of the same town, the first incompleteness result. The audience was not shaken. However, that was not true for every participant in the room. Another genius was there, who saw the consequences of Gödel's work. In the audience was a certain John von Neumann.

Gödel had to flee Austria when the Nazis came to power. A Nazi student murdered Moritz Schlick, the chairperson of the Circle of Vienna. Gödel, who was a member of this prestigious Circle, left the country. He found a job and a refuge in Princeton's Institute for Advanced Studies (IAS) and became the best friend of Albert Einstein.

To prove his theorems Gödel builds, for any complete formal system F, a statement G(F) called the Gödel sentence of F. G(F) is designed, as was Russell's paradox, to be self-referral. G(F) is the statement: *"This statement is unprovable."*

Gödel reasons by the absurd and supposes that the formal system F is complete. His idea is to attach to each statement of F a number, called the Gödel number. This allows him to translate any property of a statement of F, such as its truth or falsehood, into a property of the associated Gödel numbers. Properties of the statements can, therefore, be discovered by examining the arithmetical properties of the related numbers. Gödel builds these numbers for the self-referential statement G(F). He then applies a technique called diagonalization that Georg Cantor had developed to build a new statement. This new statement is that is neither provable nor disprovable in F, so the initial hypothesis is wrong and F cannot be complete.

For any formal system F, there is an infinity of unprovable proposals. We can take such a proposal as new axioms in a new extended formal system, as it is not contradictory with F. However, that extended system will also be incomplete and contain its unprovable statements.

A most famous example of an unprovable question had driven Georg Cantor to depression. Nobody before Gödel suspected unprovability. The continuum hypothesis that had haunted Cantor until his death was about the sizes of infinities. Is there an infinity between the infinity of Natural numbers and one of Real numbers? The question was finally proven undecidable. This happened in two steps, the first by Gödel himself, the second by Paul Cohen (1934–2007) in 1963. Cantor believed that there was no intermediary infinity, but never proved it. The continuum hypothesis had become the first problem on the list of important mathematical questions that Hilbert presented at the International Mathematical Congress in 1900.

Undecidable problems require a choice from the mathematician. Should he add the undecidable as a new axiom or not? A decision that the formal system cannot take itself. This introduces a further "subjective" aspect in mathematics. The mental space of the mathematician must decide according to his choice, with no rational reason. Both directions are possible, increasing the complexity of mathematics as a whole. It is also informing us on the mental space, its activity and complexity go beyond the limits of any formal system. So does the complexity of the universe containing mental spaces.[39]

Gödel's first theorem has separated the notion of truth in mathematics from the concept of provability. Gödel has shown that truth is not formalizable: Formal systems can express true proposals that cannot be proven true within the system.

---

[39] Contrary to (2.9) we consider here the Russian dolls with their inclusions.

For if G(F) were provable, then it would be true so that it would be unprovable in F. So it cannot be provable in F. But if G(F) is unprovable in F, then it is true. G(F) must be true and unprovable-in-the-system.

Gödel's incompleteness theorem has another fundamental consequence. Mathematical theories are of finite complexity, as we can generate them out of a finite number of axioms. But mathematics themselves are of infinite complexity. The mathematical landscape is infinitely bigger than ERP's physicalist reality.

The basic idea of Aristotle and Leibniz was dead, and it died where Leibniz wanted to anchor his idea of a truth-guiding language: mathematics. Even there, in this limited portion of the verbal world, all statements within a formal system cannot be proven true!

Not only our mathematical models of "out there" cannot be isomorphic to what they represent, but they might be subject to "internal" problems of indeterminacy if we are not very careful.

John von Neumann (1903–1957), the Hungarian mathematical genius, was a former student of Hilbert. He had assisted at the first public presentation of Gödel's first theorem and had understood its profound implications. Shortly after the conference where Gödel had presented his first theorem, he wrote to him, telling him he had deduced the second theorem, which was not presented at the conference or published yet. Both men would later meet at IAS in Princeton, where Von Neumann would design the computer architecture that we still use today. He even convinced IAS to build one of the very first electronic digital computers on the campus. He understood the capacities and the limitation of the specific formal systems that we call computer languages.

Gödel showed Leibniz's dream impossible, but that same dream would give birth to the digital computer. And the same dream would hint at why the human mental space could go beyond a digital computer.

Gödel undermined the idea of the provability of mathematical proposals in a formal system. Truth and provability became two different concepts. In the brain-centric perspective, blind spots are limits to scientific knowledge that appear because of the structure of the mental space itself. Surely not because of the fabric of "out there."

**

4.7 Algorithms and thinking

We can express some mathematical formulations as algorithms: step-by-step instructions decomposing the solving method into already known procedures. In this series of simple instructions, the algorithm allows us to solve "mechanically" the problem. We then say that we obtain algorithmically the solution. For instance, when we multiply two numbers, we apply a procedure that is a simple algorithm. An algorithm is a "method" a "how-to" procedure, detailed enough for each step to be "common basic knowledge."

An algorithm avoids us having to think each time about the meaning of what we are doing. Our mental space works algorithmically when it does not need to think, but only to apply a learned procedure. Algorithmic functioning is just one method used by the mental space. It involves only a few of the mental space capacities; it does not require, for instance, creativity, imagination, understanding or any of the attributes generated by Gödelian information in the natural world.

Solving a problem by decomposing in small steps has some similarities with reductionism. We ask about an object "what is it made of" and decompose it into the most elementary parts. An algorithm does the same but instead of decomposing an object, it decomposes a procedure or a proof.

Because of Gödel's work (and Turing in-computability that we will examine in the next paragraph), we cannot translate most mathematical functions into algorithms. There is no systematic procedure to solve them. This includes problems like finding an algorithm that would solve the truth problem. Such an algorithm would have allowed Leibniz's dream to succeed! We also know that some properties global properties of a system cannot be reached by reductionism. This analogy between reductionism and algorithms is not fortuitous. Reductionism, like algorithms, presupposes fragmentation, in "parts" or in "steps." Both these methods limit the knowledge that the mental space can gain by using them, but both are because of the verbal world and its digital approach. They do not affect the natural world.

Many non-algorithmic problems involve showing that a property is valid for an infinite set of cases, an algorithm that would examine one case after the other could then never be able to examine all the cases. It can also happen that certain steps of a solution are non-algorithmic and need a non-mechanical decision to be taken. External non-mechanical intervention is then necessary. This external intervention is the role of the natural world.

We don't even know most times if an algorithm can exist. We know several digital computer algorithms that can factor any number in a product of primes. However, computing time grows exponentially with the size of the number. We do not know if a fast algorithm[40] can exist. Our security systems (banking, credit cards, internet) assume that such a fast algorithm does not exist.

Peter Shor's algorithm is such a fast factorization algorithm. It works on quantum computers with enough Qbits. When we will have powerful enough quantum computers, it will oblige us to review our security systems.

Most of our daily activities comprise following a procedure. Those procedures avoid us having to think; we follow them step-by-step like does a computer. That is how most of us take a bath, cook a meal, do our job, and respect our daily schedule. Small

---

[40] Whose computing time would grow polynomially with the size of the number.

steps, one after the other, no thinking needed. An algorithm does not take advantage of the 100 ms nature has given us to interfere and take a conscious decision. (3.12).

Thinking becomes necessary for non-algorithmic problems when algorithms stop working. As long as we follow our expectations, and react according to our basic patterns, we have not involved the higher functions of our mental space.

Recent amazing successes of artificial intelligence show that the algorithmic domain is much larger than what we had imagined some years ago. By increasing the computing power, more complex algorithms can be used, that are not used by limited memory capacities of the brain. A new battle has started. Who will dominate knowledge, algorithms, or mental spaces? Computers can collect and store large amounts of data. The mental space, who developed out of the natural world, lags far behind their capacity to find correlations in gigantic databases. However, computers don't have a complete mental space, that would include a natural world.

In day-to-day life, we encounter many "non-mathemizable" questions, and we usually solve them by making choices. For instance, "Should I spend my holidays in the mountains or at the beach? Should I read Shakespeare or go to the cinema? Should I marry Rosemary or Nathalie?" A computer would have no basis to take such a decision for us. Our natural world must decide.

The corresponding problems in mathematics, non-algorithmic problems, are non-computable functions. We cannot decompose them in elementary steps to be treated one after the other. That situation often happens because a past step depends on a future step. Steps cannot be determined in a successive series, an appreciation can only come from a global overview. Decomposing in steps creates auto-referral loops in which computation gets stuck.

When that happens, if we only were to follow algorithms, they would leave us in the desert; from that point on, there are no more roads to follow, one needs to think. Intuition, creativity, understanding, and various feelings play their role.

Intuition and analogy can also solve some problems that can also be solved algorithmically with sufficient computing power, such as playing chess or go. Other common problems for scientists, like solving certain differential equations, are not algorithmic and need the creativity and intuition of the mathematician.

Finding an algorithmic solution to a problem may itself not be an algorithmic operation. This is most often the case and ensures computers will not replace mathematicians. We can verify a posteriori algorithmically a solution found by intuition. This happens when a mathematician proves a deep conjecture, he then formalizes his proof, and in theory, we can verify it algorithmically.

Once the thinking is done and once our way in the desert is found, it is sometimes possible to build an alternative route for others to follow. They do not have to redo the thinking. No algorithm allowed the invention of the plane, but we do not need to reinvent it each time we travel.

Algorithms do not "create" anything; they are only procedures; they cannot put out more "information" than what we have put into them. Algorithms have no imagination of their own. But creating them can sometimes require imagination, creativity, and hard work. Digital computers, and quantum computers, use algorithms.

**

4.8 The decision problem

In 1928, two years before Gödel's proof, Hilbert had proposed another conjecture critical to building his formalization program. He named that conjecture the *"Entscheidungsproblem,"* (the decision problem in German). It is more commonly known in the English-speaking world as the *"halting problem."* The halting problem raises the following question:

Can one find an algorithm that, when applied to any proposal of a formal system, would output a yes or a no answer according to whether the proposal is true or false?

The sentence asks if one could algorithmically decide if a mathematical proposal was true or false. Can an algorithm replace mathematicians? The question is a straight descendant of Leibniz's dream. Hilbert's decision problem is the mathematical version of our "truth problem." We can also formulate it by asking:

Can we mechanize proof? Can we build a machine (or an algorithm) that would tell us for any proposal if it is true or false? Like a machine detects "fake money" or like a "lie detector."

Before Gödel's theorems, we believed the question equivalent to:

Can we mechanize truth? It was the problem that Leibniz wanted to solve, and Gödel answered: no, we cannot.

Remained the question of mechanization of proof.

The definitive answer to the halting problem came eight years after Hilbert raised it in 1936. Simultaneously, but independently, Alonzo Church (1903–1995) in the US and Alan Turing (1912–1954) in the UK proved it was impossible. We cannot build such an algorithm.

Using a more familiar computer terminology, one can express Hilbert's halting problem this way:

Can one find a program that when applied to any program P as input would tell us if P, running on a computer, would halt after a finite time and give us a solution, or if P continued to run indefinitely?

This formulation justifies the English name, *"Halting problem."*

To prove that we could not build such an algorithm, Alan Turing, in his legendary 1936 paper, formalized precisely what we mean by an algorithm.

For that, he imagined a machine, now named a Turing Machine, who would, step-by-step, read and write on the squares of an infinite paper tape as instructed by an algorithm. (Figure 4.8.1.) We can imagine a Turing machine as someone who is following a set of rules; he can read and write on an unlimited roll of paper divided into squares. Every square can have only one symbol at a time. He can read only one symbol at a time and, depending on that symbol and his current "state of mind," he will write another symbol, change his "state of mind" and move to the left or right.

We can write such an instruction: (1, A, R,0, B).

It would mean: If the square that is being read contains 1, and the head is in state A, move one square to the Right and print 0 on this new square, then change the state of the head to B.

A finite set of such instructions thus defines every Turing machine: The Turing machine program. A Turing machine is what we call a "model of computation." It's a formal mathematical structure that defines what we mean by a "computation."

Turing's machine was an abstract mathematical construct. His ideas soon found implementations as physical devices in the hands of Konrad Zuse (1910–1995), and, shortly after, in those of John Von Neumann.

The physical implementation of the Turing machine shows that some questions about physical systems have uncomputable answers. As "out there," nothing can be uncomputable; this further confirms that our representations cannot be isomorphic to what they represent out there. The possibility to build a digital computer puts some doubts about ERP.

Turing's proof that the halting problem is not algorithmically realizable also uses a self-referral construction, this is how it goes:

Let's suppose "ad absurdum" that an algorithm exists that takes as input the description P of a Turing machine and some input x. Let's call this algorithm A. By definition; A does not halt if the P halts when it is fed with x and halts if the P runs forever.

Because A is an algorithm, there must be a Turing machine A' that formalizes it. One can then use A' as the description P to feed A, which means to feed the algorithm A with its description P=A'. Because of the way we have defined it, this machine does not halt if it halts, which is an obvious contradiction. Therefore, A cannot exist.

One cannot find an algorithm able to decide if a Turing machine will halt or run forever. Notice that this proof again uses self-reference, and again the result generates an unknowable blind spot. A deterministic machine with finite memory will either halt or repeat a previous state.

Just as Gödel had associated a number to any proposal, the genius of Turing was to associate a machine with an algorithm. They both build a bridge across two mental categories, create this way a self-reference, and arrive at an absurd conclusion. They can then conclude their impossibility.

The Turing machine is not the only model of computation; one now classifies these models of computation in three categories: the sequential models such as the Turing machine, the functional model, and the concurrent model. These models differ in the set of operations that are allowed. Different models may give different answers to what is and what is not computable given a set of rules. Computability is thus not an absolute notion; it depends on what model of computation one uses.

An uncomputable function for a Turing machine may become computable for another model of computation.

Remember that non-computability concerns mathematical functions, which means mental abstractions. It cannot concern out there, out there does not compute. Pretending that a physical phenomenon is non-computable is an abuse of language that mixes map and territory. Identifying what is out there with the mathematical representation can be problematic and misleading.

* *

4.9 Undecidable problems are everywhere

The Halting problem happens because we restrict ourselves to the "mechanizable" functions of the mental spaces, the verbal worlds. The halting problem is not about out there, nor is it about the natural world. Again, it is important not to confuse map and territory. Turing's answer clarifies that creativity, imagination, and understanding cannot be mechanized. These attributes require something else, other procedures are at work. As we will see in the next paragraph on Oracle machines, Turing himself was aware of the necessity of this "something else."

Thus, novelty is never algorithmic; it cannot be programmed. However, one can search algorithmically for correlations that would be hard to find using the limited capacity of the brain[41].

Roger Penrose (1931-), the great Oxford mathematician, Nobel prize 2020, and author put the Halting problem this way that excludes the possibility of solving our truth problem even in mathematics:

The perception of mathematical truth cannot be reduced to a set of mechanical rules.

I find the phrasing of this formulation that Penrose used in his conferences, particularly interesting. Truth is perceived, it is not absolute and not deduced. It is thus not "out there" nor in the

---

[41] This statement is still controversial with computer scientists. We argue here that the invention is not in the instrument but in the human brain. The microscope doesn't discover a new bacteria, the telescope doesn't discover a new planet, the algorithm doesn't discover a new correlation, The brain does it using these instruments.

verbal world. Only the natural world perceives, and it must be beyond mechanical rules.

What has been the reaction of the mathematical and physical communities after Gödel's incompleteness, followed by Turing's non-computability? Initially, the reaction seemed to be: "business as usual." A firm belief prevailed that mathematics and science could carry on much as before without bumping into exotic non-computable objects. Few noticed what the physical implementation of an algorithm could mean for the ERP approach.

Turing had shown that one can encode in physical objects not only data like in materialization (2.8) but also procedures like algorithms. On one hand, this allowed the digital computer, on the other it questioned ERP.

Incompleteness and non-computability have since shown up everywhere, in mathematics and even in physical applications. From being the exception, they are becoming the rule.

Some examples:

Hilbert submitted his 10th problem at the International Mathematical Congress in 1900; it concerned Diophantine equations. Those are polynomial equations with integer coefficients and a finite number of unknowns. Hilbert was asking if one could find a general algorithm that, given any diophantine equation, would tell us if it had a solution in integer values. The conclusive answer came from the Russian mathematician Yuri Matiyasevich (1947 -) in 1970, it was again negative.

In 1972, the same Yuri Matiyasevich proved simple questions a school student might even ask could lead directly to a rich diversity of uncomputable sets.

Marian Pour-El (1928–2009) and Ian Richards exhibited a differential equation with computable boundary conditions leading to uncomputable solutions. For certain computable initial conditions, determining the behavior of the wave equation is an undecidable problem.

Freeman Dyson (1923–2020) used uncomputability to argue for the evolutionary superiority of analog-to-digital forms of life.

Gödel and Cohen proved the undecidability of the continuum hypothesis.

Celso José da Costa (1949 -) and Francisco Doria (1945 -) explicated in 1991 examples of incompleteness and non-computability in Hamiltonian mechanics and dynamical systems theory.

In December 2009 the journal Nature published an article[42] in the news section entitled: "*Paradox at the heart of mathematics makes physics problems unanswerable*"; with the subtitle: "*Gödel's incompleteness theorems are connected to unsolvable calculations in quantum physics.*"

A long paper by Toby Cubitt, a quantum-information theorist at University College London, and his team made the news by asserting that one can reduce the problem of deciding whether a material is gapped or gapless to the Halting problem. The spectral gap is an important property of matter describing the gaps between the lowest energy levels of its electrons.

The reduction of the Gap problem to the Halting problem is such that the material will be gapped if your Turing machine halts or gapless if it runs forever. The result came as a surprise for ERP physicists. It is not surprising in the brain-centric context. For brain-centrism, the Gap problem is further evidence of the limits of our representation system; it says nothing about "out there" where there is no Halting problem.

Limits because of the mathematical structure of an otherwise well-tested model of the atomic structure of matter are limits to our third-party knowledge capacity.

---

[42] https://www.nature.com/news/paradox-at-the-heart-of-mathematics-makes-physics-problem-unanswerable-1.18983

The list could continue. Uncomputable sets and functions are no more at the periphery of mathematics, they are central mathematical questions and have made their way into mathematical descriptions of physical phenomena. In their book, Gödel's Way, Gregory Chaitin, Jose da Costa, and Francisco Doria show that there are more uncomputable functions than Real numbers.

What are the consequences for physics? As no halting problem and no indeterminacy can exist out of the mental space, this questions ERP. It also questions the universality of the algorithmic nature of our models. It opens the possibility for some mathematical representation of "out there" events to be undecidable.

As the physicist Seth Lloyd (1960 -) puts it in the abstract of his 2013 article Uncomputability and physical law[43]:

"If a physical system is capable of universal computation, then its energy gap can't be computed. At an even more fundamental level, the most concise, applicable formulation of the underlying laws of physics is uncomputable. That is, physicists are in the same boat as mathematicians: many quantities of interest can be computed, but not all."

"For example, even though the most concise formulation of the underlying laws of physics is uncomputable, short and elegant formulations of physical laws certainly exist. Not knowing in advance whether or not the quantity that one is trying to compute is uncomputable reflects the shared experience of all scientists: one never knows when the path of one's research will become impossible. The underlying uncomputability of physical law simply adds zest and danger to an already exciting quest."

Seth does not speak about the fundamental question that occupies us here. Is the ERP hypothesis still valid?

* *

---

[43] https://arxiv.org/abs/1312.4456

4.10 Oracle Machine and natural worlds

Alan Turing spent the years 1936 to 1938 at Princeton writing his Ph.D. thesis on ordinal logics under the supervision of Alonzo Church.

He had shown, in 1936, that Turing machines could not solve Hilbert's halting problem.

His 1939 paper describes what he called an *Oracle machine*. It was a Turing machine with an added "oracle" that could be accessed during the computation. The Oracle is an "external advisor" that would be consulted and activated at given steps in the calculation. That would happen when a question could not be solved algorithmically by the Turing Machine. He showed Oracle Machines would have capacities exceeding those of a Turing machine. He went on:

"We shall not go any further into the nature of this Oracle, apart from saying that it cannot be a machine."

Turing never came back to his Oracle machine after his 1939 paper. For a few years, the war would mobilize all his efforts, with the tremendous success he got in decoding the German Enigma machine.

Many authors, following the route Turing opened with his Oracle machine, have been concerned by the idea of the non-algorithmic base of the human mind. Among them Roger Penrose and Freeman Dyson.

The brilliant Italian physicist Paola Zizzi wrote in the abstract of her article The non-algorithmic side of the mind[44]:

The existence of a non-algorithmic side of the mind, conjectured by Penrose based on Gödel's first incompleteness theorem, is investigated here in terms of a quantum metalanguage.

---

[44] The non algorithmic side of the mind. https://arxiv.org/pdf/1205.1820.pdf

We suggest that, besides human ordinary thought, which can be formalized in a computable, logical language, there is another important kind of human thought, which is Turing-non-computable. This is metatought, the process of thinking about ordinary thought. Metathought can be formalized as a metalanguage, which speaks about and controls the logical language of ordinary thought.

"Hyper-computing" is the field of research aiming to conceive and to build computers strictly more powerful than Turing machines. By more powerful, we mean here able to solve problems similar to the Halting problem.

The most common opinion today is that hyper-computation cannot be physically implemented; this because of the "physical" version of the Church-Turing hypothesis. (5.14). However, the field of hypercomputing remains active on the theoretical side.

In their book, *Gödel's Way, Exploits into an undecidable world*[45], Gregory Chaitin, Newton da Costa, and Francisco Antonio Doria assert that hyper computing is just a matter of technology.

In my 2013 book, I postulated that the brain had to have along with a digital part, an analog computing component. I attributed this analog component to field computing by internal electromagnetic waves.

Miguel Nicolelis and I refined and developed this idea, showing how both computing systems complemented a hybrid analog-digital system. The idea is similar to the Turing Oracle machine in which an analog Oracle would continuously interact with the Turing machine part. In our 2015 book "*The Relativistic Brain*" we claim that the continuous interactions of the digital and the analog computation in the brain could effectively generate hyper-computation. We considered that the very existence of

---

[45] CRC Press 2012, The Netherlands

brains falsifies the Church-Turing physical hypothesis (5.11) and thus hypercomputation is physically possible.

We have proposed in this book to divide the mental space into two "worlds." The verbal world using languages and Shannon digital information and the natural world using analog Godelian information. The brain-centric description of the mental space is a consequence and supports our Relativistic Brain theory.

In an article from B. Jack Copeland and Oron Shagrir from and entitled Turing versus Gödel on Computability and the Mind, the authors compare Turing and Gödel's position on the question: is the human mind a machine? They state in the conclusion of this very interesting article:

> *In fact, Turing agreed with Gödel that the mind is more powerful than any given Turing machine. Unlike Gödel, however, Turing did not think that the mind is something different from machinery.*

* *

4.11 About computer simulations

Computer simulations are contributing fantastically to research in multiple fields, including medicine, drug research, meteorology, cosmology to name a few. We have noted how often, because we heavily rely on verbal descriptions, we confuse map and territory. Unfortunately, his also happens with "computer simulations." In this paragraph, we would like to underline that computer simulations don't simulate what is happening "out there." They rather simulate our mental space interpretations of "out there."

Shannon's information in our approach is a mental creation. It can be encoded on various subtracts but is not causal effective (3.3). To "make it effective," to do something with it one needs another mental space.

A simulation will use the algorithms and the data we feed the computer with. These are human mental space abstractions that we encode in the machine. It will respect the way our mental space has been individuated and fragmented to generate its simulation.

Let's imagine we are simulating the time evolution of an object. The first step will require measuring the characteristics of this object. The mental space interferes here because his selections, individuations, and fragmentations of "out there" are the ones that will be used.

Experimenting and measuring always involve a mental space to collect the data. Even where everything is done automatically with instruments. A mental space had to conceive the instruments and individuate what and how to measure. When a mental space is involved, the expectation mechanism, individuation, and fragmentation will intervene.

The next step is to select proper mathematical functions that will describe the "forces" acting on the object. That again is the perspective of a human mental space that separates forces from objects. One has then to transform the mathematics into an algorithm and encode them in the computer's memory.

A computer simulation of an object "out there" looks like a simulation of our mental representation of this object. The computers' mental space looks like our verbal world and is fed by our interpretations and approaches.

The "bridge" between a computer simulation and the phenomenon "out there" inevitably passes through our human mental space and is dressed up with its characteristics. It cannot be considered as a third party for us. It can only act as a tool for our mental space.

* *

4.12 Extending the explanatory landscape

Newton's approach considered time and space as absolute and pre-given. The scene on which event could happen. The observer was not on the scene, in theory, he could even not exist, it would change nothing to the description.

In special and general relativity, the speed and acceleration of the observer play a role. But not his mental characteristics. Einstein, in his discussions with the Indian poet, philosopher, and Nobel prize Rabindranath Tagore agrees ERP is a deliberate choice. It is a "religion" that allows him to do his work without having to bother about the mental space. I reproduce an excerpt in (5.4).

Brain-centrism also suggests taking into consideration aspects of the mental space. Sometimes one could explain observations or measurements by neurophysiological or mental space causes, such as expectations.

ERP science limits its search for causes to phenomena "out there." Extending the explanatory landscape to include causes in the mental space could help remove certain paradoxes and reorient research directions.

The suggestion becomes pertinent when the phenomena we study are themselves mental creations with no direct observable existence. Particle physics, aspects of cosmology, consciousness studies, the economy, education, and social sciences fall into this category.

This proposal is consistent with the idea that the "reality" we live in is mental. The brain and the mental space generate some of its aspects. We traditionally distinguish between hard and soft sciences. The latter includes psychological aspects that hard sciences will ignore.

The divergence is, however, getting blurry between observable physical bodies and objects dependent on a mental creation. For many phenomena it is not clear how much information is

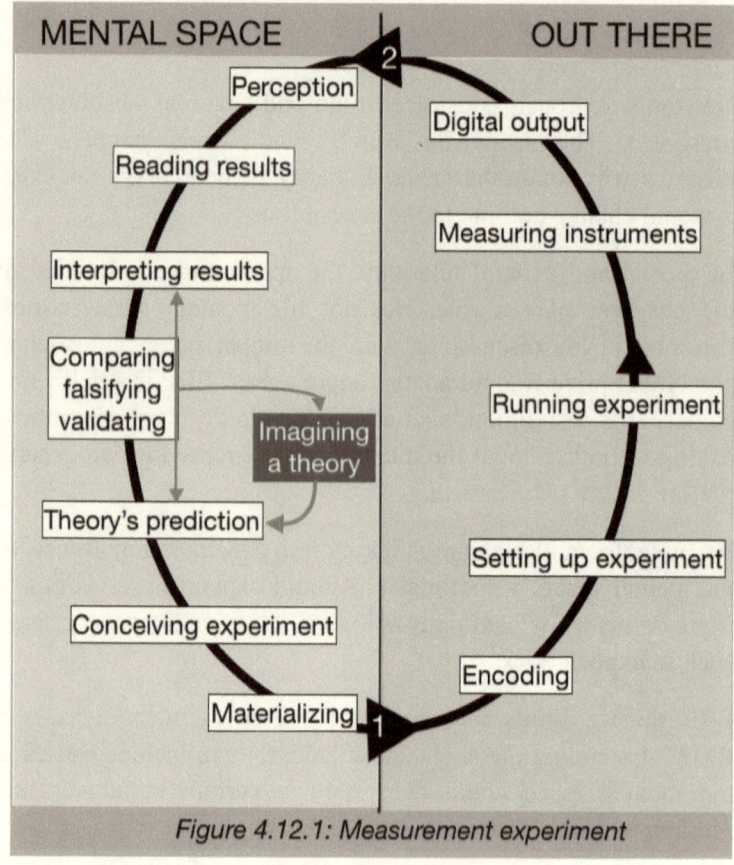

Figure 4.12.1: Measurement experiment

specific to the phenomenon, and how much the mental space originates.

In science, measurement is critical to validate a theory. It is the only bridge between "out there" and the mental space.

Let's examine the various stages we go through in a measurement experiment.

The first steps are mental. (Figure 4.12.1.) We imagine what has to be measured; we conceive and plan our experiment, its setup, and the instruments we will use. This "blueprint" is then materialized and encoded in a set-up "out there." After running the experiment, we perceive and read what the instruments have measured.

We can then compare with what the initial theory had predicted. The first and the last phases are mental. 1 and 2 in the picture represent the bridges between "out there" and the mental space where perception and materialization take place.

ERP considers only the right part of the figure (4.12.1) as an explanatory landscape. Brain-centrism suggests extending this landscape to some elements of the left part of the figure (4.12.1).

The theory that has started the experiment dictates the set-up and will create mental expectations. Often the results and their interpretation may confirm the theory's predictions. One should make sure this is not happening for reasons related to the mental space. Like a placebo effect or self-fulfilling prophecies. The expectation mechanisms, cultural censorship, or other mental effects could, for instance, introduce self-regulation and adaptation in the first and last steps of the process.

A typical example is a study of "consciousness." ERP requires explanations of the phenomenon to be found "out there."

According to panpsychism, for instance, consciousness pervades the universe and is a fundamental feature of it. Panpsychism would require new physics. According to Roger Penrose and Stuart Hameroff, we generate consciousness in cellular microtubules through quantum phenomena. They suggest we need new physics. Other thinkers believe that as we lack proper physical explanations, one should deny the existence of consciousness and consider it as an "illusion."

Brain-centrism would extend its explanatory landscape to phenomena of the mental space. Because we base the "natural world" on Gödelian information, it cannot fit in the framework of the "verbal world." Just like any other feeling. From a third-party perspective, we can only know it approximately. Our best knowledge of consciousness is first-party natural world knowledge. Thus there cannot be a "complete theory of consciousness." Even if the underlying brain "mechanism" finds a material explanation, it will say nothing about the feeling we experience. One should add that consciousness is a self-referral mental blind spot. (The camera is not on the photo.)

The initial theory (Figure 4.12.1) individuates and defines the characteristics to measure according to a fragmentation.

If this initial theory is an ERP theory, it will expect the answer originating, "out there." Brain-centrism suggests extending this expectation landscape.

For instance, medical experimentation knows the placebo effect when testing drug efficiency. It produces the required result for reasons not related to the drug itself, but because of mental space characteristics.

ERP's limited individuations have driven us to abusive approaches like Skinner's behaviorism. By focusing on "objectively measurable" criteria like behaviors, we are discounting important mental differences such as intentions. With modern new technologies, limited "objectively measurable" individuations[46] are leading us to further destructive absurdities. The domains we cited above are full of "objectively measurable" definitions concerning non-measurable mental feelings. These definitions are distorting our perception by focusing on the descriptive definition and neglecting the richness of the feeling. Brain-centrism would suggest revising the usage of defining indexes such as the GDP, Homo Economicus, IQ, …

The danger of "objectively measurable" criteria is well known in brain-centrism. We are confusing map and territory. We should remember that even when we have a common word, we don't see the same thing. Each of us experiences it differently. In the vast majority of cases, it has minimal importance. Sometimes, however, it's critical, we must consider the internal reality and not limit ourselves to the defined by description words. For instance, the judiciary tries to understand the motives and the

---

[46] These individuations are made so that the corresponding definition can be programmed on a computer, but distort the mental brain reality.

intentions of a criminal. The behavior alone doesn't assert a complete understanding.

During the first quarter of the 20th century, scientists discovered that at the microscale, when the reduction process reaches tiny scales and entities, measurements become impossible. In quantum experiments, one must account for some characteristics of the observer himself. For instance, his choice of measuring instrumentation. If he chooses a telescope or an interferometer, the results will be different. This observation will produce strange ERP interpretations of the quantum phenomena. It's the case for experiments such as Thomas Young's double-slit or Wheeler's delayed-choice experiment or Eugene Wigner's friend paradox. To explain these experimentations, physicists are led to develop unverifiable theories such as parallel universes.

* *

Summary of the concepts introduced in chapter 4:

- The verbal world is subject to self-referral paradoxes and blindspots
- These self-referral loops induce indeterminate proposals and uncomputable function
- "Out there" and the natural world do not suffer from indeterminacy or uncomputability
- Indeterminacy and uncomputability limit the capacity of the verbal world and digital computers, as not every proposal is algorithmic
- The truth problem has no solution within the verbal world, truth is a feeling of the natural world
- The brain is a hybrid computer, according to the Relativistic Brain Theory
- Computer simulations are verbal world simulations
- The mental space can affect measurement.

* * *

V: FROM INVISIBLE WORLD TO SCIENTIFIC PARADIGMS

Every epoch has its dominant background of ideas, beliefs, feelings, and knowledge. Like an invisible grid or a living collective mind, these basic ideas diffuse and shape mental spaces. Beliefs in invisible worlds seem to accompany all evolutions. Even the latest paradigms haven't eliminated them. We will inquire to understand how and why mental spaces generate invisible worlds and how and why they generate physical laws. We will examine the main paradigms that govern our thinking and the problems they cause. The two deep problems we have analyzed Descartes's dualism and Leibniz's dream are the background source of all these dominant beliefs and paradigms. Mathematics, which started in the invisible world, has played a unique role in the human mental space.

"Every kind of science, if it has only reached a certain degree of maturity, automatically becomes a part of mathematics."
David Hilbert,

5.1 Mental spaces and invisible worlds

Why is it that mental spaces need to create invisible worlds, like did our shaman? Why is it that this has happened in all known human civilizations[47]? These worlds have taken on different appearances and names. But humans, over the millennia, always needed to believe that beyond the visible there is some invisible. And that to understand the visible one must access the invisible. Rituals, prayers, or meditation but also arts, music, and love are the bridges we built to contact the invisible.

How is it that our mental space (and our natural world) needs some hidden reality and populates it with actors, stories, and myths to make sense of what is visible? How is it that religiosity

---

[47] One should ask if mammals also create invisible worlds? It is well documented that elephants have mourning and funeral rituals.

appears in the human mental space? Is dualism "a la Descartes" unavoidable? (See 4.12.)

Some humans would evoke the third information channel we have already mentioned in (3.5) to justify this belief. This third channel would put us in contact with invisible worlds and justify religiosity. The information from this third channel must be Gödelian and undetectable by third-party observations. Humans will need to interpret it in digital form to communicate it. The interpretation will generate the actors and the stories inspired by the culture of the epoch. That every tribe has different gods does not invalidate the existence of this third channel.

Once an organism moves, it has to develop sensors and a nervous system to coordinate its movements. In its evolutionary development, the brain will discover representations and expectations. It will become curious and evolve to a point where it needs to connect its representation and form a global worldview.

However, because of mental representations and causality, any worldview will always be incomplete. Explanations will be missing. There'll always be explanatory gaps. The most frequent questions are: Who created all this and why? Why do certain things happen and other things don't?

Science can answer parts of these questions. But because of its verbal, digital nature, a large part falls out of its possibilities (4.6). We can only describe a few steps backward in the causal chains.

Explanations respond to emotional needs, such as fears and curiosity. Survival is their ultimate source. They must generate understanding and fit in the global worldview to play their fill-in role. They should also contribute to the quality of predictions.

Language gave invisible worlds a powerful feeling of reality. It allowed them to develop and diversify, and their stories started spreading through generations. These narratives became the founding myths of civilizations as they answered in-temporal questions and gave meaning to otherwise unexplainable facts.

To describe his vision of "accurate knowledge," Plato presented his famous "cave allegory" in his monumental work "The Republic," in 375 BC.

Plato considers people who have lived all their lives chained to the wall of a cave. In front of them, the only visible thing is a blank wall. On the wall, shadows are projected coming from "entities." Plato imagines that fire behind the people creates the projection of the shadows. He examines what can be the "reality" of those prisoners who have only seen shadows. He explains a philosopher is a prisoner who has gotten rid of his chains and has discovered that the shadows on the wall are not reality, but only projections.

For Plato, one cannot know "reality" until one has broken his chains. The "true reality" is in a world of pure ideas. (An invisible world.) Plato is considered the founder of idealism, one of the most influential ideas of all time.

Invisible worlds are "about" the visible world they explain. They belong to a "platonic realm." They explain what we perceive from a higher but invisible perspective. They are one step higher in the abstraction ladder.

Invisible worlds were initially populated with fill-in forces like gods, spirits, monsters, angels, titans, and many mythical creatures whose wishes and behaviors explained what we observe. Their third-party anthropomorphic descriptions gave these creatures a "real" and sometimes human appearance. Although their powers largely exceeded those of humans.

Since the first known civilization, death was considered at the very heart of humans' weakness and one of the biggest mysteries. It has been the subject of the most powerful fill-in myths and motivators of humans' activities. Everybody dies, the creatures of the invisible world are immortal.

The Platonic reality is much more important than the visible reality because it creates and explains the visible one. Plato called the visible reality an illusion, the invisible reality became for him the only "existing" one. His approach fits perfectly with the expectation mechanism. We expect before we perceive. However, it is problematic. There is nothing to perceive, no feedback comes from "out there" to correct and adapt the "invisible" reality.

Over 2000 years later, Descartes tried to bridge visible and invisible realities. (3.10). Today, some neuroscientists are still trying to discover a "brain code"; to bridge the visible brain to the invisible mental space.

Materialism has tried to negate the Platonic realm which is impossible. Any materialist explanation will always be incomplete because it does not explain itself. The very existence of a materialistic explanation implies the existence of a Platonic realm in which the mental explanation exists. (5.12)

Invisible worlds appear in the mental space as the combined result of abstraction, causality, and curiosity. The mental space often uses visible entities to symbolize or personify the players of these invisible worlds. Any mental space that has developed both the natural and the verbal world will generate invisible worlds. One can thus predict that any intelligent extraterrestrial civilization will have developed them. As their source is Gödelian, every culture will translate them in different myths. Reasoning and the verbal world are not a substitute for invisible worlds. This is because once we can abstract, any explanation needs a meta-explanation. The reason is then trapped in an infinite regress causal chain and cannot explain the origins. Invisible worlds don't have the same logical constraints.

* *

5.2 The language of Gods

Once we discovered mathematics, they found their natural place in the invisible worlds. Mathematics is mental, abstract, and can explain many phenomena. Mathematics was since the beginning considered as the language of the gods. Religiosity, mathematics, and philosophy have thus a common origin as forces that populate and animate invisible worlds.

Mathematics soon occupied an important role in the invisible worlds. One could find numbers everywhere and manipulate them to get corresponding results in the material world. Long before Plato, many humans believed that what we were seeing was only an artifact, the true fabric of reality was hidden, and mathematics a way to access this realm.

When I was a child, in Cairo, I saw from my window in the distance the Gizeh Pyramids. I remember questioning, asking why. Why this gigantic, unimaginable effort to build these incredible monuments. How did these people find the energy, convince thousands, commit so many resources to these constructions? Today I have an answer. Humans do things only because they believed in invisible worlds. These worlds give meaning that matter alone does not give. Meaning is the food of our natural world, it's worth any effort.

Ancient Egyptians believed that to get the attention of the gods; one had to respect specific mathematical rules, inspired by astronomical observations. Observing and measuring the Great Bear and Orion with an instrument called a "merkhet," astronomer priests determined the proportions of every important building with astonishing accuracy. These monuments were not made of matter, but combined matter and meaning in their mental space. What made the Egyptian civilization so remarkable was not the "matter" part, but the "meaning" part.

The buildings are just the materialization of the meanings in their natural world, a sort of printout explaining their feelings and their beliefs and their invisible worlds. Just as our buildings also express what is happening in our mental space.

According to certain Egyptologists, the Gizeh plateau, on which rest the three massive pyramids and the Sphinx, is a mirror reflection of the Duat. The Duat was where the gods lived. Like the mount, Olympus was the place where the Greek Gods stayed. But the Egyptian mythology was far more subtle. The Duat, the kingdom of Osiris, is where Orion and Sirius rise just ahead of the sun at dawn on the summer solstice.

Their world was a connected whole, including life and death. Their "meanings" were profound and universal. Gods, astronomy, and mathematics were integrated into the world of matter. This weaker fragmentation gave a deeper substance to their lives that we cannot imagine today.

Thoth, the god of wisdom, was also the inventor of writing and the hieroglyphs. He wrote the "Book of the Dead." He was also the protector of writers, mathematicians, scribes, teachers, and whoever was handling knowledge and was one of the most important gods.

Egyptians already used the 365-day calendar. They determined the dates of religious festivals by watching the stars and observing conjunctions. Three thousand years BC Egyptians started using a base-ten numbering system. 1900 years BC, they also mastered fractions. Stale bread gave them antibiotics. They manipulated big numbers, in hieroglyphs, a frog represented the number 100,000. A god with his hands raised in a sign of adoration represented a million.

They associated numbers with mystical and divine meanings, as we saw, two thousand years later, in the school of Pythagoras.

Greeks started separating mathematics from the rest of the invisible world. The mathematical Platonic realm was populated with numbers and geometrical objects and had its own rules. However, mathematics has somehow maintained its attribute of "language of the gods."

The idea of "demonstrable truth" characterized Greek mathematics. Influenced by Thales, Diophantus, Euclide, Pythagoras, Demosthenes, and so many others, Greek mathematics was a specialty on its own, although it remained a part of philosophy.

Pythagoras asserted that *"all is number,"* and later Aristotles paraphrased him by saying: *"The principles of mathematics are the principles of all things."* When one entered Plato's academy, one could read: *Do not enter if you are not a geometer.*

Greeks had placed mathematics at the core of the hidden reality. Plato based his description of the Universe on the existence of only five convex regular polyhedra. Geometrical magnitudes, such as the diagonal of a square that one cannot express with rational numbers, was a sign that mathematics was a bridge to transcendence. The Babylonians had already approximated some irrational numbers such as the square root of 2. But Greek mathematicians proved these approximations can never be exact. The invisible worlds were hiding in squares and circles under their eyes, and mathematics was the bridge to access them.

The rupture of mathematics from philosophy became even subtler in Bagdad around the year 1000, discovering algebra and algorithmic solutions. They established formulas to calculate solutions for problems solved today by using linear and quadratic equations. Algebra extended arithmetic by allowing us to replace numbers with letters that could take any numerical value, forcing abstraction one level higher.

In Europe, 600 years later, the enlightenment put "reason" as the central tool to gain knowledge. It produced a further partition of mathematics from philosophy and mysticism. Galileo Galilei (1564–1642), the Italian astronomer and main initiator of the scientific revolution, wrote: *"The book of nature is written in the language of mathematics."* With thinkers like Baruch Spinoza and Denis Diderot, the enlightenment extended to the organization of society and the nature of what it means to be human.

In many regards, separating mathematics from other lines of thought remained "theoretical." The incentives to study pure mathematics, for instance, still find sources outside practical usages, and these motivations often originate with mystical thinking.

In the mental space of mathematicians, the unknown and the mysterious remain an active motivation. Most are Platonists, and many contemporary mathematicians or theoretical physicists have expressed the idea that their research was an adventure to try to *"Understand the mind of God."* This was, for instance, a favorite expression of both Einstein and Hawking. The Mind of God has even become a book under the signature of the Physicist Paul Davies with the subtitle: *The Scientific Basis for a Rational World.*

Richard Feynman expressed the same idea by saying that calculus is *"the language God talks."* Francis S. Collins is an American physician geneticist, one leader of the Human Genome Project. He wrote *The Language of God*, where he explains that science supports theism.

Leibniz had claimed that God in his universality had knowledge of all conceivable worlds, i.e., worlds who are non-contradictory. But in his wisdom had made real only the best possible world, the one with an optimum balance between good and evil. Voltaire wrote the essential little novel Candide to mock Leibniz, "best of all possible worlds."

Steven Weinberg, the famous Nobel laureate from the University of Texas, described himself as *"pretty Platonist."* He thinks the laws of nature are as real as *"the rocks in the field."* The laws seem to persist. Weinberg wrote: *"whatever the circumstance of how I look at them, and they are things about which it is possible to be wrong, as when I stub my toe on a rock I had not noticed."*

Mathematicians have excellent reasons to feel that their discipline has something eternal and divine. Once a mathematical proposal is proven true, it is true forever, leaving the feeling that it had always existed, and that we only discover it, like new territory, rather than invent it as new technology. We perhaps invent proofs, but the theorem itself has always been there, eternal. The "eternity" of mathematical truths, the rigor, and the credibility that they bring to all sciences, give them a divine origin. How do we discover them, why are certain simple conjectures so hard to prove and remain unsolved for hundreds of years, are further questions that give mathematics mystery and divinity.

Einstein famously said: "*The most incomprehensible thing about the universe is that it is comprehensible.*" For brain-centrism, our mental space has built our theories in a comprehensible way. Non-comprehensible theories never see light and are never considered. We have built our theories using the tools we understand, and that fit the constraints of our mental space.

In his recent book, Infinite Powers: *How Calculus Reveals the Secrets of the Universe*, Steven Strogatz explains, for instance, that "*calculus is as much about mysteries as it is about method.*"

The focus in this paragraph is that from the shamans up to now we need the invisible world to make sense, to understand, and to act on the visible one. We have always been Platonists. That is inevitable since we have a representation system, whose productions are not part of the "visible" world. The relation between these two worlds, Descartes's problem, remains open in an ERP perspective. Brain-centrism asserts that both worlds are mental. The progressive fragmentation of our reality these last 15,000 years have had consequences on our mental space. The natural world and the verbal world are not as integrated as they were. We lose the sense of the global, the unity, and of meaning.

* *

5.3 Physical laws

Henri Poincaré remarked that we could establish physical laws because "the Universe" was sufficiently homogeneous and regular. Without these regularities, we would have been restricted to a database of case-by-case descriptions.

However, regularities are orders and depend on mental individuations. (2.6) What appears as unusual or irregular at a time scale, such as the fall of a big meteorite, will appear as a regularity at a larger one. That would mean that our laws of physics reflect characteristics of the human mental space, not only of the universe. Extraterrestrials with a different mental space might discover regularities that escape to our eyes.

Galileo, Kepler, and Newton have taught us how to translate regularities into predictive physical laws by describing mathematically the regularities. They have taught us how to use our abstraction capacities to express what is regular.

For a physical law to make a prediction, we need to have measured initial conditions. The law is only about how the system will evolve from that point on. The law itself cannot say anything about the initial conditions. They have to be measured.

In his book: *"The Character of Physical Law,"* published in 1965, Richard Feynman explains what he means by physical laws:

The rhythm and pattern of phenomena of nature which is not apparent to the eye, but only to the eye of analysis.

Feynman stresses in his words Poincaré's regularities, who are not immediately apparent. They become apparent only with an intervention of mental space tools like mathematics, the eyes of analysis.

But Feynman's main point is we do not deduce scientific laws from fundamental theories, we reach them through layers of simplifying assumptions and approximations.

The German philosopher Rudolf Carnap[48] shared this opinion. He wrote in 1966:

Physics originally began as descriptive macro-physic narratives, containing an enormous number of empirical laws with no apparent connections. In the beginning of a science, scientists may be very proud to have discovered hundreds of laws. But, as

---

[48] We will meet again Carnap in (5.6) as an inspirer of the Vienna Circle.

the laws proliferate, they become unhappy with this state of affairs; they begin to search for underlying unifying principles.

Our scientific perception of the universe will depend on the laws of physics and on the mathematics that underlies them. As they are third-party views, physical already accounted for any changes that can happen in the universe. The laws themselves do not change; they are valid all over the universe and from the beginning to the end. As third-party perspectives, they live outside the universe itself; they transcend it. For ERP, they must belong to the associated Platonic realm.

In 1902 Poincaré wrote a small and fantastic book entitled "Science and Hypothesis." In his book, he first examines how mathematics and science develop. He then asserts we reach most scientific conclusions as a conventional consensus among scientists. Mainstream scientific knowledge is the result of convenient conventions among peers.

Poincaré showed prior hypothesis is at the base of all human knowledge. Knowledge is not "absolute"; it relies on unproven or unverifiable assumptions. At various epochs and in diverse contexts, those assumptions are different.

Conventionalism is the claim that adopted frameworks determine the physical meaning of measurements. Truth is not absolute, it will depend on two factors: measuring data and interpreting these data with conventionally accepted narratives.

Poincaré's conventionalism opposes head-on ERP. How could our theory be isomorphic to "out there" if they change and "out there" does not?

The *"focusing principle"* would rescue ERP. It is the idea that scientific laws are approximations. These approximations are getting better at each new generation of laws. Theories are focusing through their evolution on what is out there.

However, for conventionalism, as for brain-centrism, physical laws cannot be "absolute" or God-given entities. They don't come from a Platonic realm, mental spaces produce them.

Because we never observed them fail in the past, these "verbal world" proposals are crowned "physical laws."

Thomas Popper was one of the two most influential philosophers of science in the 20th century. He rejected the inductivist view for empirical falsification. He contributed to clarify the scientific method born during the enlightenments. According to Popper, we can never prove a scientific theory correct, but we can falsify it. Experiments can contradict a theory's prediction.

Empiricism requires from any narrative certain conditions for it to become a scientific theory.

—It should be logical and coherent with measurements and already accepted theories.

—It should make at least one prediction that an experiment could falsify.

—Observations or experimentations should be repeatable and validate this prediction.

Under these conditions, a scientific theory is never "absolutely true," it is waiting for a falsification.

Experimentation and measurement are thus the cornerstone of science. Shannon's capacity theorem limits the digital information flow in measurements (3.3), and our models cannot be isomorphic to "out there" (1.7). Scientific theories represent our "best-effort" knowledge at a given moment.

If the theory fulfills the first two conditions, and the third is on hold, one can speak of a scientific hypothesis. If only the first condition is met, the theory is not scientific.

In case an experiment falsifies the prediction, the narrative is no more science, and we have to fix it or replace it. We express most often scientific theories in mathematical terms, allowing predictions to be calculated. However, it is not necessarily the case, Darwin had no mathematics in his descriptions.

However, Karl Popper's "ideal" view is getting harder to follow today. This is because empirical experimentation is getting more now that we have emptied the lower branches from their fruits. Often, direct observations have become impossible for various reasons.

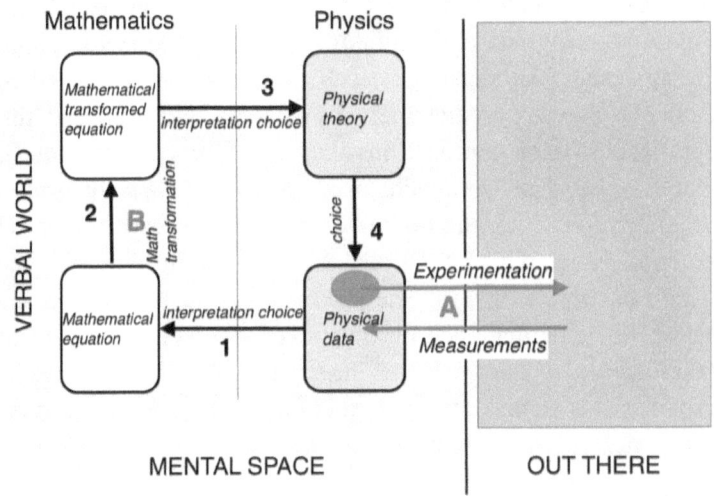

Figure 5.3.1: Mathematical transformations and physical interpretations

The conceptual background of theories often relies on non-observable entities that have only a "mathematical existence." (5.3.1) Quantum physics confronts us with situations where the simple fact of gathering information destroys the fundamental properties of the inspected entities. Particle physics requires higher and higher energies to dig into the substance of "out there." Studying living organisms, such as the living brain, forbids us to collect certain information in keeping the organism alive. Studying the origin of the universe confronts us with physical and intellectual limits.

Figure (5.3.1) represents the steps from measurements of physical data (A) to choice of the mathematics (1), followed by the mathematical transformation (2), the physical interpretation of the results (3), and the falsification or corroboration of the results[49].

At all four levels, the mental space made choices. For instance, it eliminated some mathematical results as non-physical (3), or it applied one mathematical approach rather than another (1).

In a famous lecture in 1964 at Cornell University, Richard Feynman invited the audience to imagine two objects gravitationally attracted to each other. He then asked the audience how to predict their movements and proposed three approaches. Each approach invokes a different belief about the world. The first approach used Newton's law of gravity, according to which the bodies exert a pull on each other. The second imagined a gravitational field extending through space, which the objects distort. The third applied the principle of least action. It holds that each object moves, following the path that takes the least energy in the least time. All three approaches produce the same correct prediction. They are three equally useful descriptions of how gravity works.

The same empirical results can have multiple interpretations. How to choose?

"One of the amazing characteristics of nature is this variety of interpretations schemes," Feynman concluded.

Is that a characteristic of nature? Doesn't it mean that we don't know what is going on out there? Isn't it rather a characteristic of our mental space who can produce a variety of interpretations of the data he has collected?

Leibniz, 300 years before, had made an observation similar to Feynman. He had remarked that through a finite number of points, one could trace an infinite number of continuous curves. In his mind, the points represent our measurements, and the continuous lines are our interpretations or our theories. The theory we choose will determine our worldview. Because of it, we will build pyramids, rockets, or … (5.3.2)

---

[49] Further explanations on figure (5.3.1) in paragraph 5.5.

When in 1632 Galileo wrote the *Dialogue Concerning the Two Chief World Systems*, he was in fact discussing two Feynman interpretation schemes or two Leibniz curves.

When Einstein proposed General Relativity, he was presenting a new Leibniz curve. By observing and measuring, Eddington confirmed that Einstein's interpretation was closer to the points than Newton's. The focusing principle is respected.

The focusing principle works for the mathematical predictions of our theories, they are getting more accurate at every generation. We are getting closer to the points and passing through points that had not been measured before.

It surely does not apply to the interpretations of the theories, the narratives describing "reality" are different. Just as different as the two narratives that Galileo had discussed 300 years before.

At each generation, we are accepting opposed models of the fabric of reality, based on different individuated concepts, each with its mathematics, and its language. Each generation brings its perspectives and its future possibilities and limitations, and each its mysterious unknowns, its hypothesis, and its fill-ins.

ERP will focus our attention on measurable mathematical results. Brain-centrism will welcome these advances, but will also care about the interpretations. Because the interpretations will model our worldview, not the data. (5.3.2) Interpretations will act on our collective mental space, guide and hopes, and our perspectives. They will decide what we should do tomorrow, what we should build, and where humanity is going to put its efforts. Measurements are only part of the complete picture, we need to go beyond ERP.

By extending the explanatory landscape to the mental space, brain-centrism strives for a more integrated scientific approach, closer to the full spectrum of our mental capacities and limitations.

* *

5.4 Subjective choices in the scientific method

In certain cases, empiricism will not help us choose between two theories. For instance, if all measurements would give the same result for both narratives like in Feynman example of 1964. (5.3) Or, if the data that could show a difference, is not accessible or not individuated. In particle physics, it could be because the energy required to get these data is too high. In "in vivo" neuroscience, it could be because getting it would imply destroying what we want to measure. Sometimes, because of entropy, the required information has spread out and become inaccessible.

What can then arbitrate? Let's suppose both theories are compatible with our prior knowledge. Then the only answer we can have is conventionalism.

We can always agree on measured data, how scientists choose and agree on interpretations of these data. How to choose the best curve of the figure (5.3.2)?

The mental space has to choose: What theory, what interpretation of the data to adopt? How to decide if two theories make the same predictions? And if two theories make different predictions, both validated? What if a theory makes multiple predictions and only some of them are falsifiable? How to react if an interpretation supposes that non-observable entities exist? (Arrow 3 in figure 5.3.1.) We are entering the era of post-empirical science. As we will see in this chapter, post-empirical science introduces mental aspects in traditional objectivity.

Discussing the many-world interpretation of quantum mechanics, Jim Baggott, an experienced British popular-science author, in a recent article in Aeon (2019) says:

"These theories are attractive to some few theoretical physicists and philosophers, but there is absolutely no empirical evidence for them. And, as it seems we can't experience these other universes, there will never be any evidence for them."

An extended definition of existence would allow physicists to consider that the multiple universes "exist," because they are useful to explain observed phenomena. However, other interpretations of the measurements or the mathematics (5.5.1) can also claim existence for the same reason. How to choose? Must we choose? Any choice will only rely on the mental space.

Paul Dirac in 1928 remarked the role of the mental space characteristics in the choice of a theory.

"It is not always so that theories which are equivalent are equally good, because one of them may be more suitable than the other for future developments."

Brain-centrism supports this observation that extends the causal landscape. A theory that allows future developments or has beneficial effects on the mental space is to be preferred.

Scientists use non-empirical, subjective "rules" to select one interpretation over another, or, in certain cases, one theory over another. (Arrow 1 in 5.3.1)

These "guides" are playing a crucial role in the post empiricism era. For the public, this is rather confusing. Too often the newspaper titles expose an interpretation as scientific and "absolute." These presentations erode the credibility of science at a time this credibility is essential. For many of us, certain claims are very like pre-enlightenment authoritarian truths. It is a risk that brain-centrism, when superficially understood, could stress. Being a mental creation does not reduce science to a simple opinion. It is our best possible knowledge, our most precious asset, and we should care for it and the credibility of science. Nothing should restrain scientists from speculating and hypothesizing, but we should not present speculations as science.

Let's review some guides that physicists used to make their choices.

In the Middle Ages, William of Occam, a Franciscan friar, pointed out that scientific theories ought to strive for parsimony.

Theories should avoid introducing unnecessary new hypotheses and assumptions. They should, as far as possible, rely on existing individuated entities and accepted theories. We call this principle "Occam's razor," it's an epistemological criterion, it has now become standard in science.

Leibniz has observed in his discourses that God has created the most condensed possible laws that would give the most diverse results. Gregory Chaitin remarks Leibniz had already expressed in his words, the idea of the shortest possible computer program that would generate the richest output[50]. We call a law that would generate the entire universe a *theory of everything*.

Another popular, but controversial, guide is the *"beauty of the theory."* Beauty would be an indicator that the theory is true. The idea of relating beauty to truth is old. Truth and beauty have a common source that appears in Greek and Roman philosophy, and in most religions. Beauty is "divine," mathematics is the language of the divine.

At a conference, a lady asked Bertrand Russell: *Why is it I find nature beautiful?* Russell replied with a Darwinian touch: *Madam, you'd better.* The story does not say if the lady understood Russell's answer[51].

For Paul Dirac and Richard Feynman, beauty and simplicity are guides towards truth. Physics, according to them, should prefer theories that are beautiful and simple. Simplicity here is an aesthetical benchmark, not to be confused with Occam's razor.

Einstein himself considered that *"beauty of the equations"* as a criterion, although, as we see in his dialogue with Tagore (7.5), he accepts the subjectivity of beauty, but not of truth. He did not take simplicity as a guideline. (He'd better!)

---

[50] Chaitin along with Kolmogorof and Solomonoff is the founder of the Algorithmic information theory, who study the relationship between computation and information of computably generated objects. Finding the shorts possible program is in general impossible because of Gödel/Turing.

[51] Finding Nature beautiful is positive for survival.

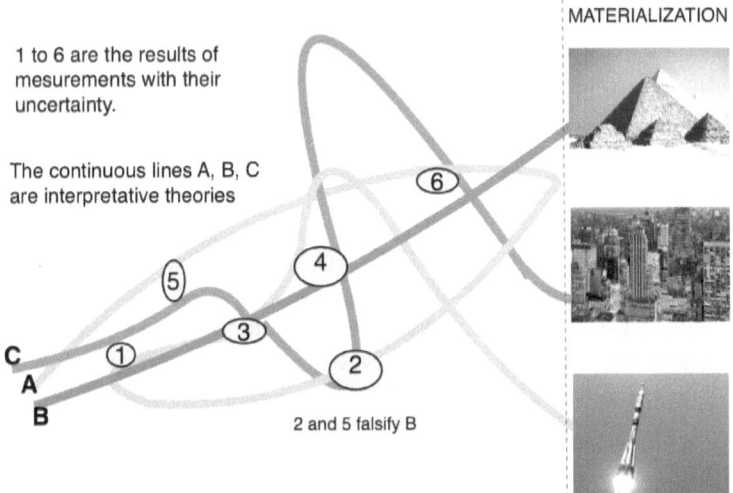

Figure 5.3.2: Leibniz metaphor. Measurement, theories, materializations

Naturalness is another modern criterion of truth that has become a force among physicists and philosophers. A theory is "natural" if its underlying parameters have a similar size. Naturalness refers to the "vague idea" that a model of physics should work without requiring some ad hoc "fine-tuning" of its parameters.

Leonard Susskind is a professor of theoretical physics at Stanford University and author of popular books. He is one founder of string theory and of the holographic principle. Susskind rejects the idea that beauty, simplicity, or symmetry can qualify the fabric of reality. He claims that conceptions of elegance are shortcuts created by our pattern-seeking brains that rarely hold up to scientific scrutiny.

Physicists have often introduced "principles" to dismiss "non-physical" results that their mathematical models would produce. (Arrow 3, figure 5.5.1.)

For instance.

The "*chronology protection conjecture*" proposed by Stephen Hawking which postulates that the laws of physics are such as to prevent time travel on all but microscopic scales.

Roger Penrose introduced the *"cosmic censorship hypothesis."* These principles eliminate certain mathematical "solutions" that appear on the map but that physicists "judge" not to exist out there.

In the last decades, a technique named "bootstrapping" implies a coherence in the laws of Nature. It assumes that the laws dictate one another through their mutual mathematical consistency. ERP physicists claim that the mathematical consistency of physical laws teaches us a lot on nature. Brain-centrism considers that this consistency teaches us about the mental space that has produced the laws.

For instance: Steven Weinberg (1933-) who won the Nobel prize in physics in 1971, had shown in 1964, that the "existence" of spin-2 particles[52] named bosons leads inevitably to general relativity. This remarkable result implies general relativity through a novel approach using only "mathematical consistency."

Bootstrapping is not without reminding Leibniz's concept of space and time. Leibniz opposed Newton's idea of absolute space and time. He considered that space and time are not so much a stage on which bodies evolve, but a consistency relation, orders outlined by the mental space:

"As for my own opinion, I have said more than once, that I hold space to be something merely relative, as time is, that I hold it to be an order of coexistence, as time is an order of successions."

In 2015, I took part at a conference in Edinburgh entitled Beauty in science organized by the late and regretted Fields medalist, Sir Michael Atiyah. Roger Penrose connected the beauty of Escher painting with the beauty of the underlying mathematics. Robert Attenborough showed us how puffer fishes had a sense of beauty. Many of the attending scientists were convinced that Beauty was a powerful guide to truth. As they also supported

---

[52] https://en.wikipedia.org/wiki/Spin_(physics)#Fermions_and_bosons

ERP and physicalism, many tried to "objectify" beauty to avoid a contradiction in their beliefs.

Brain-centrism observes that beauty is a powerful stimulating factor for the mental space. When associated with sensory representations, it generates curiosity and desire. Finding a theory beautiful, we will work on it harder. If other people find it beautiful, it will attract them. So if beauty does not say much about the truth of a theory, it has another role to play, as Dirac had observed. Beauty as a guide makes sense for mathematics. They don't require any external validation. It is perhaps the very core of mathematics. Sensing beauty, a mental space will choose to develop an aspect of the mathematical landscape rather than another one.

The very gifted physicist and fantastic blogger, my Facebook friend Sabine Hossenfelder (1976-) believes that beauty is a deceptive guide for science. She gives excellent arguments in her successful recent book (2018): "*Lost in Math: How Beauty Leads Physics Astray.*" She points out how mathematics has derived physicists to study theories who have no chance to make any experimental prediction, because of the beauty of the maths.

After reading her book, I remarked that most of the people attending the conference in Edinburgh were mathematicians. Beauty definitively guides pure maths.

* *

5.5 Sciences and mathematics

One can consider that science uses two chief guides in its search for truth: experimentation and mathematics. A and B represent them in the figure (5.3.1). Mathematics is an internal operation of the verbal world and experimentation is the bridge between the mental space and "out there."

In ERP, mathematics does not belong to the physical space, but to the associated Platonic realm (5.1). In the brain-centric perspective, it's a mental space creation and belongs to the verbal world (5.1.1). Mathematics can thus only assure that our mental manipulations are legitimate and coherent. (Arrow 2 in 5.3.1) They won't say anything about "out there" itself, nor if the mathematical tools that are used are appropriated and "correspond" to some physical operation "out there." (Arrow 1 in 5.3.1.)

However, mathematics is much more than a language. They have an internal structural coherence, a logic and a manner of reasoning, a way of organizing thoughts, that can guide us through successive steps. For instance, many theorems relate different mathematical fields, allowing us to use different representations for the same underlying object. Bridges between unrelated mathematical domains permit the transfer of underlying organization from one field to the other. By using a mathematical representation, the scientist disposes of all the tools mathematicians have prepared. This is the process illustrated by arrow 2 in the figure (5.3.1).

Newton had developed new mathematical methods to express laws of movement. Einstein found Riemann Geometry and tensor calculus off the shelf and could use these mathematical concepts and processes to establish general relativity. Murray Gell-Man found Cartan's classification of Lie groups ready for him to pick up the group su(3).

Mathematicians do not give meaning to the variables or the unknown. They pursue the most general case.

With arrow number 3, physicists will interpret the mathematical results to get a physical law. To do that, they try to make sure that the mathematics makes "physical sense" by screening the results with the guides we have described here above. (5.6). They plug-in units and, by doing that, they re-associates a physical "meaning" to the variables.

For physicists, mathematical symbols are idealizations. For instance, the most used mathematical concept is perhaps equality =. who is a high-level abstraction. To find things equal, one has to abandon a lot of differences. Mathematical equality has no "existence" in the physical world. Two physical events "out there" cannot possibly be equal. When we use equality, we refer to these abstractions of the physical phenomena (2.5), not to the phenomena itself.

Many authors have recently questioned the legitimacy of using Real numbers in physics. This includes Gregory Chaitin in his article *"How Real Are Real Numbers*? (2012), and more recently Nicolas Gisin[53], the famous Swiss quantum physicist. In the abstract of a recent paper (2018)[54] Gisin writes:

"It is usual to identify initial conditions of classical dynamical systems with mathematical real numbers. However, most real numbers contain an infinite amount of information. I argue that a finite volume of space can't contain more than a finite amount of information, hence that the mathematical real numbers are not physically relevant."

In another paper published in January 2020, in the journal Nature, and entitled *Mathematical Languages Shape Our Understanding in Physics*[55], Nicolas Gisin worries about the usage of Platonist mathematics in Physics.

As we underlined in (5.1), for our mental space to understand the universe, it has to use mental objects that, in the ERP perspective, cannot exist such as infinity, continuous functions, derivatives, and integrals, Lie groups, algebra, …

Physicists use mathematics differently than pure mathematicians. They give meanings to symbols and interpret the results of calculations through these meanings. Loading meaning into symbols leads to differences in how physicists and mathematicians interpret equations. Physicists can, for instance, eliminate certain solutions because of the guides we have

---

[53] https://en.wikipedia.org/wiki/Nicolas_Gisin

[54] https://arxiv.org/abs/1803.06824

[55] https://www.nature.com/articles/s41567-019-0748-5

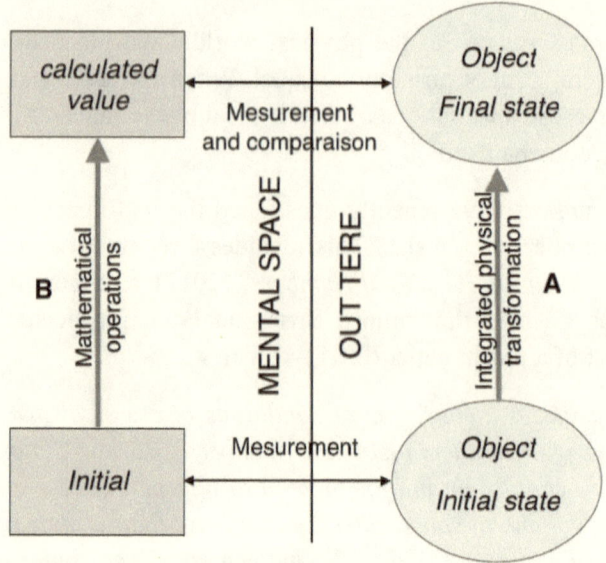

Figure 5.5 1 : Physical transformations and mental operations

mentioned and, for instance, declare certain solutions, "non-physical."

In his book published in 2014, *Our Mathematical Universe: My Quest for the Ultimate Nature of Reality*, the MIT physicist Max Tegmark claims that what we name matter, atoms, molecules, stars are all "baggage." He considers "reality" as limited to the mathematical apparatus used to describe the behavior of matter. Reality for Tegmark is a "set of abstract entities with relations between them." That would be an idealistic position, in which the Platonic associated realm would be the unique reality. He speculates that mathematics does not describe the universe, it is the universe. For him, the Platonic world of mathematics contains the visible world. This extreme Platonism reminds the Greek principle of sufficient reason as adapted by Leibniz. The claim *the universe is mathematical* is an ERP assertion. For ERP, the mathematical equations are what the universe "is." For brain-centrism, *the universe is not mathematical, our models are*. And our model cannot be isomorphic to what is out there.

Physical laws, although they can make verifiable predictions, are not isomorphic to what happens out there.

Empirical verification controls that the values predicted by the mathematical law when initial values are introduced, corresponds to final measured values (figure 5.5.1.).

The mental route (B) differs from the route "out there" (A). (B) is a mental manipulation of data got by discrete measurements. It has required fragmentations, individuations and expectations. (A) does not happen the same way. There is no fragmentation, no data, no measurement. *Planets don't know Newton's laws.* Identifying them is confusing map and territory. Getting the correct result for a transformation doesn't mean we understand how this result is produced. For our planet, it is surely not gotten by applying laws. A third-party description cannot be isomorphic to what it describes. (1.8) and (2.5).

For ERP, (A) is isomorphic to (B).

Physicists, like Nicolas Gisin, worry: *we identify initial conditions of classical dynamical systems with mathematical real numbers.*

Measurements can only produce discrete data. The mathematical manipulations use continuous methods and Real numbers that one can never measure out there. An entity out there can only "contain" finite Shannon information. (3.3). Sometimes, the mathematical model uses complex numbers that have no "reality" out there. In other cases particles move backwards in time, an event that has never been observed.

Conventionalism is harder to accept with mathematics. How can the Pythagoras theorem on square triangles result from a conventional agreement? How could it possibly depend on us humans?

Most pure mathematicians are "Platonists," they tend to believe that mathematics has an independent "existence" in an "ideal" realm. (5.1). The work of the mathematician is to explore this realm and to discover its components.

Mathematical objects are defined by description. The "existence" of a mathematical object is equivalent to its definition. The definition is not the mental representation of a mathematical object, it's the object itself. The definition must not contradict the axioms chosen for that mathematical theory. For a mathematical object, the map is the territory. Nothing has to be measured out there, and no interpretation is necessary. Existence for a mathematical object does not depend on any other source of information, be it sensory or of emotional origin. If one accepts that the definition is non-contradictory, one accepts that the object exists.

However, in their work, mathematicians do attach images and feelings to mathematical objects to trigger their intuition and their imagination. As Kuhn explained, it is only in the context of justification, when their work is communicated, that any reference other than formal must be abandoned. The mathematical notion of existence put it aside of science.

As we examined in (5.1), either as ERP would suggest one considers mathematics belongs to a Platonic realm. Brain-centrism asserts, it belongs to the verbal world.

Gödel's indeterminacy theorems show that, within a formal system, one cannot prove its coherence and that formal systems are incomplete (4.5). The mathematical landscape, including all choices of axioms, is of infinite complexity. Mathematicians conventionally chose what territories of this landscape are interesting to explore. They chose in relation to already known mathematics, but also to humans' intuition. Some territories are too far for our intuition to make valuable advances. They consider interesting the parts of the landscape where they can make discoveries. (6.9).

We formalize proofs of theorems for publication and conventional acceptance. Some mathematical proofs take years to get accepted; because of how complex they are and how much time to go through the proof.

Conventionalism thus also applies to mathematics.

An interesting observation, partially due to Newton da Costa and Francisco Doria, could make clear how intuition interferes and why aliens could have different mathematics.

The Jordan Curve theorem says that a simple plane closed curve, such as a circle, with no self-intersection divides the plane into two disconnected regions. The set of points inside the circle and the points outside the circle. This seems intuitively obvious to our mental space. Rigorous mathematical proof is astonishingly complicated. Some extraterrestrials with a different mental space might not find the Jordan Curve theorem obvious. They may not have the same geometrical intuition as we do because of their different brain physiology. However, we both will agree on formal proof. This is the power of mathematics, once the terms defined and the axioms agreed upon, one can communicate, they are no more matters of opinion or interpretation.

Terrestrial mathematicians have probably concentrated on the mathematics for which humans have good intuitions, rather than regions of the unlimited mathematical landscape where our brains leave us with no exciting visions.

The structure of our mental space does not leave us any choice. We have to collect and analyze data through our sensory system. We have to interpret these data to build mental models and interpretations. As Dirac observed, these interpretations will condition future developments. Because of the difference in nature of (A) and (B) our models give sometimes unacceptable results. Physicists need guiding principles such as those we described in here above to filter and select the results that make physical sense.

**

5.6 The Circle of Vienna

Years after Poincaré had studied and proposed conventionalism, after Frege's, Russell's and Hilbert's tentative and Gödel's and Turing's results, the dream of "absolute" truth that Aristoteles and Leibniz had been pursuing, was still not dead. It might never die.

Figure 5.9.1: Trurh dancing

Over the centuries, two entities have been dancing together. The dance is the search of truth; the music is obsessive and beautiful. The two dancers have, however, a different idea of what their dance is looking for. They realize they can both take advantage of the other's movements. The first dancer seeks truth by dancing rigor, formalism, logic, and mathematics. His movements are digital, they come from the verbal world. His partner, the second dancer, knows truth cannot be hidden there and searches for it through the harmony of a global analog dreams in the natural world. The first dancer can show his immense accomplishments, he has to his credit knowledge and technology. The second dancer believes that these accomplishments only happened because of her. She senses he has produced those things not because of his movements, but because she inspires him. If she were to follow him, both would get lost. The dance would stop. He is looking into more and more details, hoping to find the secret hidden deep into matter, and by creating useful applications, she is searching *the mind of God*. Descartes vainly tried to marry them. Materialism pretends he is dancing alone.

We humans with our hybrid digital analog brain, with our double mental space, we humans are dancing our lives.

Thomas Kuhn (1922–1986) was, with Karl Popper, one of the two most significant philosophers of science of the last century. Kuhn is known for having introduced the notion of "paradigm" and paradigm shift. He exposed his ideas, in 1963, in his very controversial book: *"The Structure of Scientific Revolutions."*

The Guardian newspaper rated it among the 100 most influential books of the 20th century. However, when it was first published, it raised a lot of criticism.

To understand why it raised so much criticism, it is necessary to situate the philosophical context in which Kuhn's ideas developed. Astonishingly, this context is related to our central thesis: how the mental space produces our realities.

Among the many scientists that fled Europe to avoid the Nazis, a large group of them came from Austria. In the 1920s and '30s, the *"Circle of Vienna"* was one of the intellectual world centers. The German physicist and philosopher Moritz Schlick (1882–1936) had started the circle in 1923. It lasted until Schlick's death in 1936, when a Nazi student assassinated him at the entrance steps of Vienna's university.

Vienna had become, at the beginning of last century, home to many great thinkers among which, notably, Ludwig Boltzmann, Paul Ehrenfest, Wolfgang Pauli, Ernst Mach, Otto Neurath, Herbert Feigl; but also Sigmund Freud, Leon Tolstoi, Leon Trotsky, Joseph Stalin.

The Circle of Vienna was initially named the "Ernst Mach society," to honor the giant Austrian physicist who left his name to the speed of sound, and was an inspirer of the young Albert Einstein.

Ludwig Wittgenstein's *Tractatus logico-philosophicus* (1921) was the first "Bible" of the Circle, although Wittgenstein himself never took part in any of the meetings.

After publishing his book, Wittgenstein was convinced he had solved all problems of philosophy. These problems he believed arose from a misunderstanding of the logic of our language. The Tractatus asserted that in order to determine the truth of a narrative, one has to compare it with "reality." That comparison was the work of natural sciences: *"The totality of true propositions is the total natural science."*

So after the publication of the Tractatus, Wittgenstein abandoned philosophy and became a schoolteacher in his hometown of Vienna. But in 1929, he changed his mind; he returned to Cambridge, where he had studied with Bertrand Russell, to meet Frank Ramsey (1903–1930). The young Ramsey had visited twice with Wittgenstein in Vienna and had persuaded him there was a flaw in the Tractatus. Ramsey claimed substantially that logical relations are not entities of the world "out there," but connections between mental narratives.

The underlying philosophy of the Circle was called *"logical positivism."* Rationalism and empiricism inspired it. For logical positivism, information derives from sensory experience, and mathematics is the exclusive form of all *certain* knowledge. Logical positivism was an ERP approach, and if initially Moritz Schlick was open to Kant's ideas on phenomenology, he then changed his position. The book published in 1928 by Rudolf Carnap entitled: *"The logical structure of the world,"* was probably at the source of Schlick's change.

Carnap, that we have already met in (5.2), was a German-speaking philosopher, who became a leading and very influential member of the Circle. He is still today considered as one of the major philosophers of the last century. Carnap, together with Wittgenstein, recognizes that many philosophical problems are simply the result of the misuse of language. Some of these problems simply disappear when one acknowledges they concern choices between different linguistic frameworks. Following the traces of Aristotles and Leibniz, Carnap considers that the logical analysis of language is the key instrument in resolving philosophical problems.

The Circle made some references to the work of Henri Poincaré, Gottlieb Frege, Bertrand Russell, and Karl Popper. It sowed a foremost consideration for natural science, mathematics, and logic. Empiricism and "objectivity" were not subject to discussion. The Circle was pluralistic and committed to the ideals of the Enlightenment. Under the influence of Wittgenstein and even more Carnap, the movement wanted to prevent confusion due to language and denounced illogical and unverifiable claims, in the tradition of Aristotle, Leibniz and Hilbert. (5.4).

Niels Bohr, Albert Einstein, and Oskar Morgenstern are believed to have occasionally taken part in some of the Circle's encounters. Kurt Gödel, who then lived in Vienna as a Ph.D. student, became a junior member of the circle.

The meetings were held at the University of Vienna or at Cafe Central, one of the famous Viennese coffee shops. The key participants could have a seat around the central table and be allowed to speak one after the other. Gödel, who was a junior member, could only sit on the chairs arranged along the walls and could only silently listen. The movement also had a branch in Berlin where certain meetings were organized. Doing the '30s, the Circle had gained an international reputation among the world's intellectual elite.

Logical positivism distinguishes, in a Cartesian way, between two kinds of "objects": the "observable" ones, like "this chair," and the theoretical ones like "this electron." Contrary to Descartes's vision, and because logical positivism was an ERP perspective, both types of objects have a physical existence.

The tentative of the Circle was to relate theoretical objects to observable ones using logical considerations. Logic would then be the bridge between theory and direct observation in the spirit of Leibniz mathematical structuring of language. Like Descartes's, this tentative ultimately failed at explaining how theory can relate to observation. In this *"Observable versus Theoretical"* divide, a logical bridge would have its two ends on the same side, in the verbal world, which is astonishing for a bridge! Logic can only relate a representation to another representation, both ends of any logical or mathematical bridge belong to the mental space's verbal world. Nothing except observation or measurement can connect it to "out there."

Isn't it a curious irony that it was a member of the logical positivist Vienna Circle, Kurt Gödel, who showed that logical systems containing arithmetic could not prove their consistency and thus could not lead us to the truth? And even more interesting, two deep questions, both initiated 2500 years before by Aristotles, after following different routes, would finally meet up in Vienna. (4.5 and 4.9)[56].

When Kuhn's book was published, logical positivism had expanded worldwide, and become one of the main philosophical positions among scientists. The idea Gödel had developed in the 1930s had in fact left a very little impact outside a few specialized mathematicians and logicians. Logical positivism had continued to spread worldwide after the war, and the scientific-philosophical background was "a priori" not so appropriate to welcome Kuhn's ideas.

This is because Kuhn explains one has to distinguish two different contexts. The context in which we make a discovery, from the context in which the discovery is justified and communicated. For him, discovering and communicating are two aspects of the scientific work. In our term here, discovering is a natural world analog activity that has the be translated to the verbal world and structured to be communicated.

The context of discovery, that logical positivism "de facto" rejected, is non-algorithmic, it involves intuition and creativity, it is often motivated by the feeling one rejects something commonly accepted. The context of justification and communication is algorithmic, rigorous, formalized and logic. Natural world versus verbal world. Logical positivists were only open to consider the context of justification. For them, rational arguments and logic must understand the scientific enterprise. Kuhn gave its importance to the context of discovery and the "irrational" (non-algorithmic) aspects of thinking. He situated it at the heart of science, in the discovery phase. He got vividly criticized by the scientific establishment.

---

[56] Even more astonishing is that the divide between theoretical and observable objects remaining opened, it was artificially solved by introducing an extended definition of existence, that introduced mental operations in ERP objectivity without saying it!

The brain-centrist description of our mental space supports Kuhn's asymmetry between discovery and justification. It recognizes the central role played by emotions in our interpretations, our creativity, and our intuition. It asserts that logic and rationality were developed in a later phase and that algorithmic functioning hardly contributes to imagination. We handle the discovery phase in a first-person perspective, as all natural world activity. In the scientific process, third party, rational and communicable narratives only appear later, in the context of justification. The only possible "bridge" between theoretical and observable objects is experimentation. For brain-centrism, experimentation is only made possible because the mind can act on the body through emotions and expectations, as described in (3.11). Brain-centrism also insists on the importance of individuation, expectation and interpretation in the fragmentation, the selection, the filtering and interpreting the measured data.

This is clear in mathematics, discovering and presenting are different activities.

* *

5.7 Paradigm shifts

The most important idea detailed in Kuhn's book is the concept of *"paradigm shift."* It details how science advances, alternating phrases of profound novelties and phases of digging up the details. Both phases are necessary, like both our dancers lead the way.

A scientific paradigm, as a high-level mental abstraction, can only be "vaguely" defined intensionally. However, Kuhn described two components:

1- A set of fundamental theoretical hypotheses accepted at that time.

2- A set of typical scientific problems that are solved with these theoretical assumptions.

A paradigm is not a scientific theory, it concerns the scientific environment and its procedures; it is rather an agreement a la Poincaré between powerful scientists on how future researchers in the field should proceed, which problems are "interesting" to tackle, what are the methods to solve these problems, what the solutions should look like. Such presuppositions are meta-scientific, they are not science.

A paradigm is like a scientific program, a guide to what, when, and how things should be done and presented to be good science. This set of shared assumptions, beliefs, and values imposes itself on the scientific community. For scientists, it defines "normal" science at a period. Under the reign of a paradigm, conjectures or theories that do not fit in the limits the paradigm has fixed are discarded. Their authors have to abandon or are rejected. A paradigm facilitates science communication, scientific education programs, and above all financing of research programs. Under the umbrella of a paradigm, "truth" can be presented as more "objective," research directions are clear and one can just sit down and do the work of discovering.

What logical positivists disliked the most in Kuhn's approach to scientific knowledge is that it affirms that science rests on unproven assumptions and beliefs. Absolute rigor, absolute truth, absolute objectivity, no intervention of subjective choices, was considered the very heart of sciences. In the ERP "zeitgeist" of 1963, rigor, precision, and denial of any trace of subjectivity seemed essential. Science had gained an aura of credibility that distinguishes it from any other type of narrative, because of this objective rigor and precision. That was a character that the scientific community was not ready to see discarded.

Although Kuhn's ideas became mainstream, science is still today confronted with the dilemma. Rigor and truth give science its aura, its acceptance in the public, and its financial resources. However, it is based on assumptions and intuitions. Fundamental science is like creativity, it cannot be pre-planned. So the dance continues.

Kuhn's ideas on scientific truth appear as a natural extension of Poincaré's conventionalism, expressed already 60 years before. (5.7).

For Kuhn, "normal" science is the type of science done under the reign of a paradigm. This is what he says about normal science in his book: *Normal science does not aim at novelties of facts or theory, and when it is successful, it finds none.*

- Wandering in the labs of a university or a technology company, one can see thousands of talented researchers. Young Ph.D.'s full of hope, high levels, very busy technicians, and confident professors, all are doing normal science. Many of them are not conscious they are living under the reign of a paradigm. Many of them are convinced that what they are doing is the way science must be done. They feel they are working to build our future. We organize the entire system for them to believe in the importance of what they are doing and to keep them dependent on the paradigm. The organization runs through the internal hierarchy, the publication system, the financing system, authority, diplomas.

- From time to time you meet a denier, a woman, or a man who has put a foot outside of the paradigm, a dreamer seeking the mind of God; the system will squash him. From time to time, one of them escapes and will make history. I just reviewed the scientists cited in this book. With no exception they were all deniers, none had been doing normal science.

- Deniers are excluded because elements of the paradigm itself are non-negotiable. According to Kuhn, a period of normal science can last until too many anomalies of the paradigm have accumulated. Too many facts derived either from the theory or from measurements do not fit within the limits of the ruling paradigm. In the beginning, there are not enough anomalies, and we simply ignore the isolated ones that appear. We must save the paradigm as long as possible. For instance,

- Aristotelian physics dictated the Earth was at the center of the universe. It failed to explain the retrograde motion of the planets. Claudius Ptolemy saved this paradigm when too many questions became burning. If the planets are revolving around the Earth, then why do they sometimes pause and reverse direction? Ptolemy modeled this anomaly using epicycles, as he described in his Almagest. This "patch" to Aristotle's work lasted another 1500 years. Until the invention of the telescope, Copernicus, Galileo, and the enlightenment.

- Under Newton's paradigm, one supposed that "ether" filled the entire universe and made up a stationary frame of reference. Ether had to be impermeable to electromagnetic waves but completely permeable to matter. That strange or paradoxical property was "ignored" until Einstein removed the idea of either altogether and changed our narrative on gravity.

- Einstein added a "cosmological constant" to his General Relativity equations to maintain the Universe's size invariable. He later called this addition his biggest mistake.

- Today in quantum electrodynamics (QED) the vacuum energy is estimated at 10^{113} joules per cubic meter. Using the upper limit of the cosmological constant, the vacuum energy of free space has been estimated to be 10^{-9} joules per cubic meter. This huge discrepancy is known as the cosmological constant problem. The problem remains as weirdness physicists have to live with; it might require a paradigm shift.

- Lord Kelvin, in 1884, first suggested the existence of "dark matter" to explain why galaxies rotation did not follow Newton's laws. It is accounting for approximately 85% of the total matter in the universe if our laws are correct. In 1906, Henri Poincaré in "The Milky Way and Theory of Gases" used the expression "dark matter" in discussing Kelvin's work. In 1980, a paper by Vera Rubin and Kent Ford showed most galaxies must contain about six times as much dark than visible mass, according to our theories or gravity. Scientists widely recognized the need for dark matter as a major unsolved problem in astronomy. They developed several hypotheses to explain dark matter and dark energy and rescue our theories.

- The dominating model of the brain is still today based on computationalism. (5.12). This model of normal neuroscience is the paradigm that will allow scientists access to publication and financing. Each region has a function; eventually, each neuron has a function! Miguel has shown that this reductionist paradigm does not hold, and the brain being a complex system one must also consider it as a "continuity," as a whole to understand it.

- A famous debate opposed Einstein and Walther Ritz about the right way to understand classical electrodynamics. As two electrons have both negative charges, they will repel one another. The question was: does an electron feel the effect of his electromagnetic field? Is there a self-referral effect, as we have observed in various examples? Both answers, yes and no, lead to problems where infinity shows up, or energy loss becomes unexplainable. Feynman, who was Wheeler's assistant, proposed, as had done Ritz, to eliminate the idea of a wave. In the Wheeler-Feynman theory, particles respond to both the past and the future behavior of one another. But the problem has up to now no consensual ERP solution. That is not astonishing. The brain-centric proposes to extend the explanatory landscape (4.12). Mental spaces are the ones who generate self-referral problems, not "out there." (4.12.1).

According to Kuhn, when the anomalies with a paradigm exceed a certain threshold, everything stops. The confidence in the paradigm breaks down, and we stop producing normal science.

A scientific revolution period can then start. It will last until we build enough confidence in a new paradigm. And a new cycle can start based on the new consensual agreement that becomes the new main scientific paradigm.

The accepted paradigm explains the previous anomalies. It requires some reorganization of the field. We have to examine how the systems fit together under the new perspective.

Sometimes the reorganization extends beyond strict science and will bring modifications to the dominant worldview, as has been the case during the enlightenment.

* *

5.8 The Laplace deterministic Universe

Although the concepts of determinism, predictability, and reversibility are closely related, they are different. Determinism is a third-party proposal about a system, thus an ERP property of the system independent from any observer. Predictability refers to the capacity of an observer to model the future evolution of a system. Reversibility qualifies the independence of a mental model regarding the time parameter. By abuse of language, reversibility is used by ERP to qualify the system itself. Here, one has to suppose it is an isolated system.

ERP and brain-centrism have different views and attach different meanings to these three words.

The Greek philosopher Leucippus (5th century BC) was the teacher of Democritus. They have together developed the theory of philosophical atomism.

Leucippus is known for having stated a dogma that remains familiar today under the name of the "principle of sufficient reason."

"Nothing occurs randomly, but everything occurs for a reason and by necessity."

The picture proposed by Leucippus is that of a world of causal entailments, understandable by the human brain. The past events determine the future events. Randomness, things that happen with no cause, is excluded. However, Aristotle, in his Physics and Metaphysics, contradicts Leucippus's dogma. He accepts that there are "accidents" caused by "chance." Chance becomes itself a cause, an uncaused or self-caused cause like randomness.

Must of us remain today convinced that everything happens for a reason, that any event has a cause. Our mental space needs to find reasons to make sense of its representations. For brain-centrism, this need developed because of the mental space's internal organization necessities. Remember that causality is the brain function allowing energy-saving reorganization. (3.9). Learning and understanding happen by generating and changing causal chains of representations and feelings to form meaningful networks. Any event must fit in such a network, even if one has to employ an ad hoc fill-in.

Leibniz nearly 2,000 years after Leucippus would develop similar ideas:

"Everything proceeds mathematically ... if someone could have a sufficient insight into the inner parts of things, and in addition had remembrance and intelligence enough to consider all the circumstances and take them into account, he would be a prophet and see the future in the present as in a mirror."

But the champion of a fully deterministic universe is the French polymath Pierre Simon de Laplace (1749–1827).

Laplace worked in engineering, mathematics, statistics, physics, and astronomy. He was one of the first scientists to postulate the existence of black holes. Someone had told Napoleon that Laplace's book, *"The System of the World,"* contained no mention of the name of God. When both men met, Napoleon questioned:

"M. Laplace, they tell me you have written these large volumes on the system of the universe, and have never even mentioned its Creator."

The story says that Laplace answered bluntly:

"I did not need that hypothesis."

The "Laplacian universe" has remained the symbol of a deterministic universe. Once the initial conditions are established, no room is left for randomness or God's, or anybody else's intervention in the succession of events.

In 1814 Laplace presented *"Laplace's Demon."* If the demon knows the position and the momentum of every atom in the universe, he can calculate these values for any moment in the past and the future. The universe is not only deterministic, it is also fully predictable. However, the demon has to know its state at some moment.

In his book dated 1814, *"A Philosophical Essay on Probabilities."* Laplace, extending explains:

"We may regard the present state of the universe as the effect of its past and the cause of its future. An intellect which at a certain moment would know all forces that set nature in motion, and all positions of all items of which nature is composed, if this intellect were also vast enough to submit these data to analysis, it would embrace, in a single formula, the movements of the greatest bodies of the universe and those of the tiniest atom; for such an intellect nothing would be uncertain and the future just like the past would be present before its eyes."

Laplace claimed he did not need the god hypothesis, that the entire universe could be self-explained.

Paul Davies remarked that Laplace's demon would have a hard job in measuring his initial conditions and that quantum mechanics did not yet exist:

If the demon overlooked the gravitational force of a single electron at the edge of the observable universe, then his prediction for the motion of a given molecule of air in your living room would be rendered completely uncertain after only 12 intermolecular collisions. This arresting example reveals how exquisitely sensitive to error predicting the future can be... The real absurdity in Laplace's statement is its implicit reliance on physical Platonism extrapolated to a staggering degree...

A certain number of physicists, including David Deutsch, have rescued Laplace's Demon from quantum indeterminacy and true randomness by adopting the "Many-world interpretation[57]" of quantum mechanics.

The Laplacian universe would now be a Laplacian Multiverse.

Laplace Demon, and the many-world hypotheses, assume ERP that brain-centrism denies. The demon himself would live in the associated Platonic realm and would have to store his data there.

* *

5.9 Physicalism

Max Planck, at the beginning of the last century, was trying to clarify the role and methods of physics and of science. He was aware of the distortions that could bring human perception. His idea, in the line of Aristoteles and Leibniz, was to make science as indisputable as mathematics, he declared:

"The role of physics and science, generally, is to obtain a complete separation between reality and individuality of the brain. It is thus the role of physics to build a world stranger to consciousness, where consciousness has no role to play."

Lord Kelvin expressed the same idea in 1883:

" ... when you can measure what you are speaking about, and express it in numbers, you know something about it; but when you cannot measure it, when you cannot express it in numbers, your knowledge is of a meager and unsatisfactory kind; it may be the beginning of knowledge, but you have scarcely in your thoughts advanced to the stage of science, whatever the matter may be."

This struggle for "objectivity" that most scientists since Galileo and Newton have adopted paves our road from an analog world

---

[57] The many world interpretation holds that there are many worlds which exist in parallel at the same space and time as our own. The existence of the other worlds makes it possible to remove randomness and action at a distance from quantum theory and thus from all physics.

to digital one, but it eliminates from science investigations an important part of what makes up our human lives.

A few years later, Erwin Schrödinger would write:

"The material world has only been constructed at the price of taking the self, that is, mind, out of it, removing it; the mind is not part of it..."

By removing the mental space from the science narratives, physicalism generates third-party perspectives that concentrate on matter.

Craig Weinberg writes:

Materialism ultimately predicts that consciousness is impossible, which is why we are left with materially impossible terms like illusion and information to plug the gigantic hole it creates.

This third-party view is, however, also a mental space production with constraints aimed at eliminating biases. We call it a semi-third party view. It considers mental properties as "illusions," artifacts of neural activity. This is an oxymoron, the illusory artifact is claiming that he is an illusory artifact, just as in the liar's paradox. (4.5)

Physicalism is a one substance view of reality. It's the thesis that everything is physical. It is opposed to dualism a two-substance view as Descartes had proposed. (3.10) Physicalists don't deny that the universe may contain items that don't appear to be physical, they consider such items are physical or supervene on the physical. For our usage, physicalism will stand for materialism.

Physicalism is today the mainstream paradigm in science.

What does it entail to adopt physicalism?

- Physicalism is an ontological monism and thus an ERP paradigm, which means it considers we perceive what is out there. It does not distinguish between map and territory. It doesn't explain the binding problem. (1.3)

- Physicalism must consider that its laws, who are valid for the entire universe, are not part of that universe to avoid an infinite regress. They must belong to an associated Platonic realm whose description is not part of physics. (5.1).

- Physicalism entails the mind and body problem. It has to consider our feelings, our inner life, and consciousness as illusions.

- Physicalism must limit its explanatory landscape to out there. Causes must belong on the right part of the figure (4.12.1). Causality itself is physical, and physicists explain it by energy exchanges. However, energy uses causality to be explained.

- Physicalism obliges scientists who adopt it to introduce and justify concepts like time, entropy, order, information, or randomness as properties of "out there." It also obliges them to extend the concept of existence.

- Physicalism must explain limitations to measurements, and knowledge by properties of "out there." It cannot consider certain limitations as properties of the mental space. (4.12)

- Physicalism denies the mental space and only considers the underlying brain and its activity. As it is not measurable, it denies Gödelian information. The only type of information is Shannon information that becomes a physical property "out there."

- Physicalism treats consciousness as an illusional passive spectator, registering an action and believing it has generated it. It does not explain the content of conscious thoughts or why experience accompanies these physical processes. The nature of such an "illusion" is not explainable in physical terms. Illusion is relative to something that we do not consider an illusion. No "physical illusions" can exist out there at the ground level. One has to presume some base-level mental perspective to begin with. An illusion needs a non-physical-mental space to "exist."

Physicalism is thus a paradigm whose consequences in interpreting reality differ from brain-centrism.

For instance, under physicalism:

- One could pursue research on "brain uploading" as all information is Shannon subtract independent.

- Computationalism could seem plausible. We will see in the next two paragraphs that even under physicalism computationalism is dubious.

- General Artificial Intelligence could be a plausible research aim.

- A general brain code allowing mind direct reading would also be plausible.

- Quantum physics will assert that true randomness exists out there.

In his book *Something Deeply Hidden*, Sean Carroll, a famous physicist from Caltech, declares:

"As far as we currently know, quantum mechanics isn't just an approximation to the truth; it is the truth."

Carroll thus identifies maps and territories. One can understand his position if one remembers a particle is equivalent to its Shannon information content. Just as a mathematical object "is" its definition, a particle has no "observable existence." Quantum mechanics is then "reality" rather than "about reality." But it is purely mental reality.

Objects and behaviors defined or predicted by mathematics create mental expectations. As they are not observable, these expectations have no way to be updated by sensory inputs. This is the reason Carroll says quantum physics is true and not an approximation of truth, in the same way as mathematical proposals are true. However, his remark is confusing. What is true in quantum physics is the mathematics, not our description of "out there."

To explain itself, matter needs non-matter. Descartes frontier between body and mind can only exist if it does not exist. Materialism and idealism cannot explain themselves without involving each other.

*

The most famous argument against physicalism is Frank Jackson's (1986) *The knowledge argument*.

Mary is a brilliant neuroscientist who has learned all the known physics about color and vision. She has read all the books and studied all the classes, but Mary is color blind. The argument then goes that by some surgical intervention, Mary gets cured, opens her eyes, and sees the blue sky. She learns something new to her: what it is like to see blue! She has an unfamiliar experience that she could not have learned before by description. This phenomenal experience is blueness! The argument is that something is missing in the physicalist description of the world.

Physicalists argue Mary has only learned a new behavior, which is: how to react to color knowledge by acquaintance, but no new knowledge about the color itself.

The brain-centric view explains that her representation of color blue got updated by sensory information from her eyes that was not there before. Her physical description of blue as an electromagnetic wave of 480 nm has not changed, but she has discovered a new feeling and related it to the physical description. Physicalists do not estimate that Mary has learned anything. As they consider, feelings are illusionary emerging properties. For brain-centrism, she learned nothing learnable by description but learned a lot by acquaintance.

* *

5.10 Computationalism is a bad brain model

Neuroscience is unfortunately plagued by a false paradigm that has been driving research these last 30 years.

The Atlantic published in July 2019 an article entitled: *"The Human Brain Project[58] hasn't lived up to its promise. Ten years ago, a neuroscientist said that within a decade he could simulate a human brain. Spoiler: It didn't happen."* These are the first lines of the article:

On July 22, 2009, the neuroscientist Henry Markram walked onstage at the TED Global Conference in Oxford, England, and told the audience that he was going to simulate the human brain, in all its staggering complexity, in a computer. His goals were lofty: "It's perhaps to understand perception, to understand reality, and perhaps even to understand physical reality." His timeline was ambitious: "We can do it within 10 years, and if we do succeed, we will send to TED, in 10 years, a hologram to talk to you."

There are many reasons Markram claim could never happen, most of these reasons are detailed in our book with Miguel Nicolelis: *The Relativistic Brain, how it works and why it cannot be simulated in a Turing Machine.* (2015). For our purpose here, I would like to concentrate on two arguments, who are hidden presuppositions made by Markram and his followers. These arguments went unnoticed by the funders and deciders who supported the project. They are crucial for brain-centrism.

The first is an ERP assumption. *By copying or simulating all observable components of an entity, we get a similar entity.* As ERP neglect non-measurable characteristics, it will handle an organism as it handles a mechanism. The basic method used by experiments similar to the human brain project is to *"reverse engineer the brain"* as we do it for machines. One simulates the components in the computer and animates them with equations approximately modeling their behaviors. This idea voluntarily ignores the fundamental differences between an organism and a mechanism. An organism cannot be reversed engineered. It has never been engineered in the first place. It has evolved. It also ignores that the reductionist method does not apply to complex

---

[58] https://www.humanbrainproject.eu/en/brain-simulation/

adaptive systems as we saw in (3.8). And finally, it ignores the main characteristic of a brain: to be alive.

The second assumption is "computationalism." Computationalism is a pure ERP vision. It considers that the fragmentation our mental space operates to gain knowledge is effective "out there," that planet's orbit because they follow Newton's laws.

The computationalism hypothesis asserts that our brain "is" or can be modeled by a digital computer.

There are some ways that the computer-brain metaphor makes sense. We can interpret the firing of a neuron as a digital signal. Measuring this information can be interpreted as digitally coded information.

In The Relativistic Brain, we reminded the principles governing the behavior of neuronal ensembles.[59] These confirmed principles should have by themselves long ago raised serious doubts on the validity of the computationalism metaphor.

The *distributed principle* asserts that all behaviors generated by complex animal brains like ours depend on the coordinated work of populations of neurons, distributed across multiple brains structures. The *multitasking principle* shows that the electrical activity generated by individual neurons can contribute to multiple neural ensembles simultaneously. The *neural degeneracy principle* posits that a distinct combination of neurons can produce a given behavioral outcome, at distinct moments in time. Multiple neural ensembles can yield the same behavioral outcome at different moments in time. In fact, some evidence suggests that the same combination of neurons is never repeated to produce the same movement.

These principles are the basis of neuroplasticity, a ground-level adaptation that computers cannot achieve.

---

[59] These principle put forward by Miguel Nicolelis and his team have been confirmed by decades of neuroscience experimentation. A complete description of these principles can be found in Miguel Nicolelis' book: Beyond Boundaries (2010:)

Computationalism has, however, been the dominant view in neuroscience these last 30 years. As a paradigm that fits perfectly with ERP and physicalism, it has given birth to large research programs with considerable funding, including the Human brain project cited here above. Many other programs to simulate the brain on a computer are ongoing, searches for neural codes, brain emulations, brain reading, brain uploading, artificial general intelligence, intelligent robots, and lie detectors. Computationalism has also been the basic assumption in many science fiction novels, films, and futuristic predictions. It has conditioned the dreams on several generations, although it is a false assumption.

A well-known article on transhumanism[60] one can read:

"Mind-uploading thus aims to free us from any material substrate, be it organic or non-organic." Preliminary research on mind uploading, also known as whole brain emulation, is well on its way. Academic publications, like the "International Journal of Machine Consciousness" devout entire issues to the topic.

Entrepreneurs like Elon Musk, companies like Google, many neuroscientists, philosophers like Nick Bostrom support transhumanism. Governments all over the planet have joined in the race to simulate or emulate the human brain on a digital computer, based on the false computational paradigm.

How could this false hypothesis attract so much public attention, and why have companies and governments invested billions in related research programs?

The eminent physicist Lee Smolin had declared:

Neuroscience is a fabulous area to work in, ripe for great discoveries. ... But that is a field which is as bedeviled by outdated metaphysical baggage as physics is. Specifically, the antiquated idea that any physical system that responds to and

---

[60] Antonina Kulchitskaya in her article entitled: Mind-Uploading: The Hard Problem of Transhumanism

processes information is isomorphic to a digital programmable computer is holding back progress.

Stanislas Dehaene (1965-) a famous French neuroscientist and author considers computationalism model of intelligence wrong. He asserts that unlike computers, *"neurons not only tolerate noise but even amplify it to help generate novel solutions to complex problems."*

Historically, the idea of imitating our body is deeply grounded in our psychology, imitation is the basis of learning. Think of ancient statues, dolls, fairies, demons, ventriloquists exhibiting their puppets and other mental or material objects we produce imitating humans. (See also 7.1)

Legends or mechanical marvels, human history all over the world describes moving and speaking imitations of humans. Something deeply buried in our mental space pushes us to "imitate" ourselves, using the technologies of the moment. Psychologists would perhaps say that it is a way to know ourselves or to believe we are as powerful creators as gods. When digital computers were discovered, they immediately raised the question of imitating our intelligence.

* *

5.11 Why is computationalism wrong?

The conceptual origin of computationalism is in the so-called Church-Turing thesis. The Church-Turing Thesis (CTT) asserts that any function that is *"effectively computable"* is computable by a Turing machine. CTT is a hypothesis about computable functions (4.7). It states that a function on the natural numbers is computable by a human being following an algorithm, ignoring resource limitations, if and only if it is computable by a Turing machine. A Turing machine can do what a human does when he is following step-by-step an algorithm.

Before this precise definition of computable function, mathematicians often used the informal term *"effectively calculable*" to describe functions that are computable by paper-and-pencil methods. In (4.8), we saw that a brain does not find its solutions algorithmically and can, for instance, solve uncomputable problems, for which algorithms cannot exist. Turing himself exhibited uncomputable problems like the halting problem (4.7).

An un-careful interpretation of the CTT has led to the extended idea to CTTP that asserts that whatever happens in nature could be simulated by a Turing Machine, including whatever happens in the brain. CTTP is the designation for the Physical Church-Turing hypothesis. CTTP is not what the original Church-Turing hypothesis claims! CTT is a thesis about mathematical mental maps, not about natural territories (3.2). It makes no sense to speak of the computability of a natural phenomenon unless one confuses map and territory. It can only make sense of discussing the computability of our mathematical representations. Nature does not compute, planets do know Newton's laws. Computability is about arrow 2 in the figure (5.5.1), not about "out there." Computationalism is a mistake related to the confusion ERP makes between maps and territory. This confusion favored CTTP, a physical hypothesis extending the mathematical CTT hypothesis. CTTP asserts no physical process exists a Turing Machine cannot simulate. The same way as many believed that no object heavier than the air could fly. (Although birds are heavier than air.)

Let's cite the philosopher and computability expert Jack Copeland[61]:

"An error, which, unfortunately, is common in modern writing on computability and the brain, is to hold that Turing's results somehow entail that the brain, and indeed any biological or physical system whatever, can be simulated by a Turing machine. The Church-Turing thesis does not entail that the brain (or the mind, or consciousness) can be modeled by a Turing machine program, not even in conjunction with the belief that

---

[61] https://en.wikipedia.org/wiki/Jack_Copeland

the brain (or mind, etc.) is scientifically explicable, or exhibits a systematic pattern of responses to the environment."

The major consequence of the CTTP hypothesis is that there can be no computation method that can go beyond the universal Turing Machine. And thus that the brain must be a computer.

A Turing machine can only have a verbal world using Shannon's information. Physicalism would not recognize the role of the natural world in our thinking processes. This denial leads physicalists to computationalism. All along with this book, we have stressed the critical functions of the natural world in cognitive processes.

In his article: *On the Church-Turing Thesis*, Germano D'Abramo analyzes the limits of the Church-Turing thesis.

The Church-Turing Thesis (CTT) is a crucial assumption for the proper setting and for the proper understanding of many fundamental unsolvability results in mathematics and logic. Outside mathematics CTT, in his physical/strong form (CTTP) has paramount implications in computability theory but also in the foundation of physics and knowability, in simulability of physical processes and existence of physical processes not simulable by Turing Machines.

In his paper, he gives an example of a procedure satisfying Church Turing asymptotically but that cannot be implemented on a Turing machine.

The great Argentinian mathematician Gregory Chaitin in his book *Gödel's Way* (with the Brazilian physicists Da Costa and Doria), denies that Church Turing should be extended and that it is only technical questions that are stopping us from building hyper computers.

This is what Kurt Gödel had to say in the '60s about the philosophical implications of his theorems on the brain:

Nothing has recently changed in my results or their philosophical consequences but certain misconceptions have

been put aside. My theorems only show that the mechanization of mathematics, i.e., the elimination of the spirit of the entities is impossible if one wishes to have a satisfying system and foundations for mathematics. I have not shown that there are mathematical questions who are undecidable for the human mind, but only that there can be no machine (or blind formal system) that can decide all the questions of number theory...
Letter to Leon Rappaport dated August 2, 1962.

Quantum computers cannot solve the computability problem and save computationalism. The brain is no more a quantum computer than it is a classic digital one.

We can simulate any quantum system with a classical system, but with an exponential slowdown. But everything computable via a quantum system is computable via a classical system, just not necessarily as fast.

Just as digital computers, quantum computers cannot simulate the brain.

* * *

VI: UNKNOWABLES

Science establishes constraints on what is possible or not "out there," which limits are because of "out there," and which ones are due to the structure of our mental space itself? We will question here what is the nature of the limits to scientific knowledge? We will also ask ourselves how the mental space can sometimes overcome these limits by using non-algorithmic capacities such as intuition or creativity. Self-references and blind spots scatter various regions of the mental space and are responsible for many of our knowledge gaps.

6.1 Undefinable words, antinomy

The most fundamental and exciting words are undefinable by description. They are subjects of discussion for philosophers and poets, but also scientists and computer wizards! Words like space, time, love, gravity, intelligence, trust, …

Some words are undefinable because they refer to the mental natural world's activity. We can only learn them by an acquaintance, the information that specifies them is Gödelian analog and subtract dependent. An abstraction into Shannon subtracts independent digital information will always be fragmental and incomplete.

Translating a feeling into a narrative moves us to another world. From the first person, we step into the cold and anonymous world of the third person description. The new world is governed by other rules, where causality replaces analogy. Instead of feeling, we now count and measure. A globally colored impression gets dissected and its pieces are arranged in prefabricated boxes. Now everything must have a cause and be arranged according to logical imposed rules. We are now counting the notes instead of enjoying the music.

An abstract word will slice our dearest feelings and threw away the pieces that don't fit in standard boxes. Once language got hold of them, we will hardly recognize them. These undefinable words express feelings like hate, shame, love, desire, proudness, integrity... Or they sensations like disgust, warmth, joy, pain, tenderness...

Other words are undefinable by description because they are "primitive." They enter any narrative that would define them, creating a self-referral loop. These are words like space, time, thinking, understanding, or consciousness.

Some words are antinomies, we have attached to them a feeling that does not correspond to their descriptive definition. Antinomy arises when a word refers both to a feeling and to a concept defined by description. Our mental space has then difficulties matching both first- and third-party perspectives. This difficulty is not astonishing as one is "about" the other, the same word belongs to two abstraction levels. At each level, it means something else.

The words of a scientific language are third-party perspectives. They need to be precise and univocal. We can only define them by description. Because the verbal world is subject to indeterminacy and undecidability and the universe is not (4.6, 4.8), science has limited description capacities. The scientific description tool is corrupt, and science makes sure not to consider the source of the corruption: the mental space. It pretends to avoid it by using a semi-third party language.

Words in the ordinary language come associated with other significant symbols that can change the meaning the listener will give them. They come with a context, a tone of voice, a rhythm, and the body language that will color them. Understanding ordinary language happens because the words we hear evoke a network of feelings and associated concepts that we named Net(R)[62]. (1.7) The word itself includes no meaning. The meaning is attributed by the listener. It will depend on how the listener develops Net(R). Empathy contributes to understanding. We feel what the other experiences because we have a similar natural world.

---

[62] *The logical context representation network*

We do not require exact descriptive definitions to speak the ordinary language. Being too formal might destroy the sincerity and spontaneity of the communication by introducing other elements in Net(R).

The imprecision of ordinary language is the key to the richness and subtlety of human communication. What our verbal world calls imprecision, our mental space interprets as a second layer of information that extends the range of Net(R) in the natural world.

We can say things in a large variety of different ways. One can use a metaphor, allusions, irony, humor, special environments, formal settings, specialized languages... Speech is just an ingredient of the communication between humans. With a similar culture and background knowledge, we will need fewer words to communicate.

The individuation generated by our experienced reality, the one of our natural world, doesn't correspond with the individuations proposed by the language. The way we attach feelings and analog images to a word will vary from one individual to the other. We all have our own "interpretations." Natural representations don't superpose with the descriptive definitions we learn later. A first party and third party cannot match. They never match when we speak about feelings.

We cannot show abstract concepts and feelings, one has to define them intensionally as a third party. This verbal "objectivization" suggests that these abstract concepts or feelings exist "out there." Because everybody uses them, it contributes to the ERP belief that these conceptual objects exist as properties of nature. The verbal definition of feelings imposes itself on the mental space. The "objective definition" contradicts the first-person information that our natural world feels.

This contradiction creates antinomy. An internal conflict develops between a descriptive definition and what is felt. This can be very upsetting for young adults. They do not understand the struggle between the natural and the verbal. Many conclude that they are not "normal," because they do not experience what the verbal definition wants to impose, something must be wrong with them.

If a person completely adapts to the verbal definitions, she gets mechanized. Verbal definitions of feelings are useful for robots who mimic humans, not for us that experience humanity. There is an enormous difference between being and imitating that we call the *imitation fallacy*. (7.1)

Time and space are examples of words that generate antinomy. Their meaning develops in the natural world by acquaintance at an early stage in life. Later we learn the descriptive definition but continue experiencing them as natural feelings. For instance, we don't perceive time like what our clock is saying. Our feeling is contextual, not the watch.

Immanuel Kant employed the word antinomy in his "*Critique of pure reason*" (1781). Antinomic proposals for him are contradictions that result from our attempts to conceive the nature of transcendent reality. He explains antinomies relate to four situations:—Space, time and the universe—the whole as the sum of its parts, free will and causality, the necessity of a god.

Kant resolved antinomies by distinguishing two categories "phenomena" (third-person perspectives, as we called them) and "noumena" (first-person perspectives). He insisted we can never know the noumena of things from a third person's perspective. Brain-centrism confirms Kant's view.

Digital computers use a formal and univocal language. They don't have a natural world and ancestral feelings to rely on and thus ignore antinomy. For computers to be useful tools, we need them to be reversible and thus avoid any ambiguity. Computer scientists use words describing human attributes to qualify the computer's functions. Like intelligence, memory, understanding, creativity, learning, deciding. These words mean something else for the mental space. (2.1) That can sometimes be very confusing. Humans relate words to a network of feelings that we called

Net(R) that will generate meaning. Computers have no natural world, they don't have meaning or understanding.

When a computer scientist says that a computer "understands," we unwillingly assimilate this to human understanding. And activate an entire network of connected representations and feelings in our mental space. A network that belongs to the word understanding in the human sense, but does not apply to the computer's "understanding." This network will generate a wrong image of what the computer is doing! Because of these wrong images, we end up believing that computers can be, for instance, intelligent even more intelligent than humans.

One can say *computers have a better memory than humans*. However, we are comparing two non-comparable things. A computer cannot memorize like human does, and humans don't use digital addresses to retrieve their memories.

The sentence *computers have better memory than humans* is only true if we use the computer definition of memory.

A lady at a conference given by Jurgen Schmidhuber (1963-), one of the leading tenants of artificial intelligence, asked if computers could have feelings. She wanted to know how the machine could feel "desire," for instance. Schmidhuber answered, straight forward:

"It's very simple; you give me your definition of desire, I will program it into the computer. I can program any feeling that you can define for me!"

Schmidhuber is correct by saying that if you can define something by description, you can also program it. You can program a robot to have the behavior you have described, and it will appear to be in love. Each time you launch the program, a typical imitation fallacy starts (7.1). The source of the fallacy is the belief that a descriptive definition carries meaning. The definition will trigger a behavior of the computer, it cannot trigger any feeling or meaning.

Computationalism would imply that everything in the universe is univocally definable by description (5.10). Computationalism has the same flaws as behaviorism.

Brain-centrism asserts that Schmidhuber's individuation of a feeling is wrong. It claims that any comparison of two events cannot be limited to the visible behaviors, but should include the whole procedure that generates the behavior.

I am always surprised when a machine expresses words that show an underlying feeling. When you reach Geneva airport by train, some recording machine says: "*Welcome to Geneva airport.*" The machine uses the same words as humans do, but these words are empty, no feeling, and no personal intention, no person is behind. No person, no meaning!

Some want us to believe that we can replace humans with imitated humans; others realize certain words cannot be defined by description. One has to learn them by experiencing them as humans do.

**

6.2 Truth and belief

Aristoteles had proposed his three laws of logic:

1. Identity: Whatever is, is 2. The law of contradiction, an assertion cannot be both true and false, and 3. The *excluded middle principle* we examined in (4.3).

These laws express the fundamental characteristics of reasoning. The excluded middle holds that any assertion about an existent object or event must be true or false; no middle ground exists for truth. The laws of logic did not, for Aristoteles, derive from anything deeper and applied to every existent "thing" in the universe. The aim of logic for Aristoteles and the peripatetic, his followers, was the elaboration of a system that "*allows humans to investigate, classify, and evaluate suitable forms of reasoning.*"

For ERP, logic must then be out of the universe and belong to a Platonic realm.

I remember long discussions we had among high school students on the origin of logic. Where did it come from, how does it apply to anything in the universe? It seemed miraculous, something happening in our head that applies everywhere. How did the universe learn logic? Was logic a "God-given" compass for us to understand in the universe and find "truth"?

Brain-centrism describes logic as a set of "orders" about mental representations. (2.6) Logical rules should govern our verbal representations and their relationships for our brain to become a good regulator. (2.8)

So how can it be that logic is universal wherever we observe, "out there"? How is it that the universe "is logic," if logic is a mental space attribute?

Animals were learning logical behaviors by trial and error, well before Aristoteles abstracted them as logical principles. We developed and adapted to the surrounding universe by learning. Logical principles are the very first regularity evolved animals learned by acquisition to survive. Over billions of years, survival has pushed organisms to adapt. Our brain developed, modeled by these necessities, dictated by the universe that it had to learn. The learning was Gödelian and manifested by physical transformations. "Out there" one can find no Shannon information or digital regulators; things are not symbols, they are.

We learned analogy and causality from the universe. Although they don't exist out there. What "exists out there" is a succession of situations out of which we abstracted analogy and later causality. That is how our brain formed networks of connected representation that we called Net(R) in (3.11). These networks reflect our elementary logical principle.

Logic is a mental space attribute, if the universe applies it, it's because we learned it from him. More precisely: *it's what we abstracted from what we learned from him.*

No mysterious God-given compass is guiding us to some "absolute truth."

But now it becomes far more interesting: Aristotelian logic does not always work.

Until the beginning of the last century, this "God-given" compass pointing to truth, seemed to apply everywhere. As often with knowledge, the universe showed us a problem.

Here it was the observed spectrum of black-body[63] radiation did not match with the theory at higher frequencies. The problem is the *ultraviolet catastrophe*. One could have called it the *logic catastrophe*. And as always, this problem has led us to fantastic new discoveries, quantum physics.

Max Planck proposed a law by which radiation could only change its energy in a minimal increment. This minimal increase was proportional to the frequency of the electromagnetic wave associated with the radiation. That law would solve the black body radiation problem.

We know 1905 was the *annus mirabilis* for Albert Einstein. The first of the four revolutionary papers he published that year was about the photoelectric effect. In that paper, he proposed that light be emitted, propagated, and absorbed in energy quanta, and photons are the carriers of electromagnetic radiation. Contrary to Plank's wave approach to radiation, Einstein proposed a particle approach.

For Einstein, energy is emitted through minimal packs, quanta, and not continuously in waves. Planck, Einstein, de Broglie, Bohr, Heisenberg, Schrödinger and many others led physics to accept the wave-particle duality. In quantum mechanics, one can

---

[63] A black body is an object that absorbs all the radiation incident upon it

describe every particle or quantum entity either as a particle or as a wave.

This contradicts Aristoteles' three laws of logic.

The wave-particle duality became the first of a long list of "strange" quantum properties. Although quantum mechanics is the most well-tested theory in physics and its applications are uncountable, its meaning or interpretation is still not clear. Physicists have proposed many interpretations. Each interpretation describes a different universe. No experiment will tell us which one to choose. No experiment will answer the question, "what is reality?"

In quantum mechanics, because of the wave-particle duality, one cannot state any absolute mathematical facts about quantum entities. All facts about measurable properties are stochastic.

This contradicts ERP knowledge and, of course, the three Aristoteles laws of logic.

If our primitive brains learned logical principles from the environment, we had never directly confronted quantum mechanical phenomena. They only happen at a microscale. Quantum properties seem therefore strange and their interpretation does not fit in the worldviews our mental space has built up to now.

However, quantum properties, if they contradict the classical ERP views, fit with brain-centrism, which proposes much simpler interpretations of quantum weirdness.[64]

We have got an answer to our truth question, the verbal world cannot insure its coherence. We cannot find an algorithm that will tell us if a proposal is true or false. The verbal world can therefore not totally and isomorphically describe out there. ERP is a false assumption. Physicists, sticking with ERP, are trapped in a contradiction that they try to solve by attributing properties to the universe.

---

[64] See part 2 of Brain-centric.

The role of science is to isolate belief from knowledge. Isolate and be conscious of our assumptions and premises. Brain-centrism is a tentative in this direction.

*

As we can never prove things true, we rely on beliefs. In science, Poincaré's conventionalism replaces absolute truth with a common belief of experts that is temporarily adopted as a paradigm. (5.7)

In mathematics, absolute truth can exist because the objects belong to the verbal world. They never refer to feelings or to "out there." A theorem will remain true, whatever mental space considers it if it can adopt the language, the premises, and the conventions.

In everyday life, beliefs model our thoughts. Beliefs, and not the truth, guide humans to think what they think and do what they do. Even if we "a posteriori" justify our behaviors with logical arguments. This remark is an illustration of our human condition. We have no possible guidance in our verbal world. And we have lost the primitive guidance that offered our natural world.

If we refuse to navigate from one belief to another in political, social unrest, and smoking unfairness, what should we do? Aristoteles, Descartes, Leibniz, and so many others tried to solve the problem. They did not manage to, but they made it clearer at every step: this major human problem compromises our long-term survival.

Why then is "truth" so important? An answer lies in Ashby's good model theorem (1.7). A model of "out there" not sufficiently isomorphic would inevitably lead us to disappear. The entire perspective of science is to establish good models.

With today's knowledge, it is crucial to understand how beliefs form in our mental spaces. Why we cannot avoid them, and how the verbal world can manipulate them.

Beliefs are verbal world expressions closely and strongly related to feeling. They most often concern non-visible events: future or past events, invisible world, ... Often their verbal expression doesn't appear as a direct expression of the feeling. For instance, one could believe in the existence of flying saucers and it would be based on a deep desire to discover something new. Psychology has been closely studying how certain beliefs are associated with primal emotions.

One distinguishes levels of beliefs. The higher the level of belief for a narrative, the higher the confidence in its truth. This idea has allowed statisticians to apply the Bayes theorem to express the degree of belief in probabilistic terms[65]. They then study how this probability would evolve when additional information is released. The network of parameters supporting a belief is called a *Bayesian network*.

A belief can grow to the point of denying the information of one's sensory system (1.4) when it is supported by a group. That often happens in religious fanatic movements. The belief takes over any thinking capacity. The entire world is perceived through the eyes and interpretations of this belief.

Narratives are false if, when characterizing a subject, they express something in disharmony with data collected on the territory. The territory can be this case be an event "out there" or inner emotions. Most of our knowledge comes through narratives other people have produced. It is thus important to trust them and to understand their belief system.

Science has here a critical role as our ultimate certitude. We must all be sure that science and research are doing their best effort to search for truth. Undermining the social trust in science by false claims or bad journalism is a social catastrophe.

Lying, spreading false information to others, compromises their evaluations and survival capacities. Lies have killed more people than all other diseases together.

---

[65] Bayesian statistical methods use Bayes' theorem to compute and update probabilities after obtaining new data

To express truth is a decision everyone must take personally. It requires some personal work. Finding our inner balance is difficult but crucial, as declared Albert Einstein:

As for the search for truth, I know from my own painful searching, with its many blind alleys, how hard it is to take a reliable step, be it ever so small, towards the understanding of what is truly important.

George Orwell would add concerning human societies:

"The further a society drifts from the truth, the more it will hate those that speak it."

The most common model of distributed truth in history has been "authoritarian truth." That a truth imposed on human societies by religious or political authorities. We enforced authoritarian truth for millennia, leaving people with no other choice than to believe narratives imposed by authorities.

Often authoritarian narratives manipulate fears and desires to coerce, "Beliefs." Social conformity pressure would further strengthen them. Exclusion or even execution was the fate reserved for the deniers.

Authorities created various narratives, references, and words, such as "sacred," or "blasphemy." They claimed tablets or books to be God's words. They instituted inquisitions and severe punishments for non-believers. Very serious questions, such as the creation of the universe, or the nature of human beings, have been overused.

Kings realized that authoritarian truth was a precious tool to control the loyalty of their people and keep them working. Authoritarian societies restrained imagination, doubts, and creativity. Societies couldn't make any progress. The belief was much more powerful than observation and occupied the mental spaces.

However, to be efficient, the supreme authorities, dictating the authoritarian truth, had always to belong to the invisible world.

In that way, nobody could question them. Pharaohs were living gods, we crowned most kings with God's blessings.

Most narratives used fears such as eternal life in hell or having one's body eaten by monsters or any other punishment that would happen after death, after leaving the visible world.

The goal of authoritarian truth was to control people's thinking. In many authoritarian cultures, rational thinking itself became the criminal, books were burned and declared useless.

We hardly realize today how profound have been the changes that occurred during the enlightenment age in the 17th and 18th centuries.

By imposing reason and experimentation instead of authority and belief, enlightenment opened our minds to a world of hope, dignity, research, and discovery.

It allowed us to use the full potentialities of our mental space, rid of ancestral fears. But the authoritarian temptation, that allows powerful people to reign, is still ready to recover its prerogatives. Today it uses a combination of old techniques and new technologies.

* *

6.3 Existence in physics

In physics during the last 100 years, the notion of "existence" has been under high pressure to develop. After the enlightenment, experimentation was critical. "Existence" became roughly synonymous with "being observable."

The wave theory of light outs a limit to optical observation when the sizes of the object observed decrease. Optical microscopy cannot distinguish objects separated by less than about half the wavelength of the light used to enlighten them. We often refer to this limit as the diffraction barrier. The maximum resolution in visible light corresponds to distances of around 200 nm. The smaller the entity, the smaller the wavelength of the photons has to be. The photon's energy is inversely proportional to its wavelength, which means one needs higher and higher energies to probe the smallest scales.

We had already abandoned the idea of direct observation in many sciences studying the past, for instance. For the very small, a physicist would adopt the same idea. If an object is not observable, let's observe the traces he has left.

An object would then exist under the condition that his existence explains the traces.

Physicists would agree that if one can observe the consequences of the "traces" of a mathematically predicted entity, this entity also exists "out there." (3.2).

A physical entity is then said to exist if it is useful to explain an observation. This type of existence is a combination of mathematics and physical observation. It introduces some blur in Descartes's clear separation between mental and physical. It is a step in direction of brain-centrism as it recognizes the necessity of brain activity to individuate existence.

This definition not only applies at the quantum level for smaller entities. Cosmology uses it to determine past events such as the Big Bang or inflation.

All the recent discoveries in particle physics exist according to this extended definition. A coherent mathematical theory predicted in 1964 the existence of a massive scalar boson with zero spin, no electric charge, and no color charge. In 2012 at CERN near Geneva, a particle with the expected properties was discovered. Two experiments the ATLAS and CMS experiments at the Large Hadron Collider (LHC) indirectly confirmed the Higgs boson. The properties of the discovered particle matched the expected properties of the Higgs boson. In 2013, Peter Higgs and François Englert, were awarded the Nobel Prize in Physics for their theoretical predictions.

Dark energy "exists" because we can observe an accelerated expansion of the universe that the unobservable dark matter would produce. Dark matter "exists" because it can explain the geometrical configuration of galaxies and preserve our theory of gravity. So dark energy and dark matter "exist," not because of an observation, but because they can explain an observation (2020).

In 1933, Swiss astronomer Fritz Zwicky, while observing the motion of galaxies in the Coma Cluster, began wondering what kept them together. There wasn't enough mass to keep the galaxies from flying apart. Zwicky proposed that some kind of dark matter provided cohesion. But since he had no evidence, his theory was quickly dismissed.

Then, in 1968, the American astronomer Vera Rubin made a similar discovery. She was studying the Andromeda Galaxy at Kitt Peak Observatory in the mountains of southern Arizona when she came across something that puzzled her. Rubin was examining Andromeda's rotation curve or the speed at which the stars around the center rotate, and realized that the stars on the outer edges moved at the same rate as those at the interior, violating Newton's laws of motion. This meant there was more matter in the galaxy than was detectable.

A young and brilliant professor of theoretical physics, Claudia de Rham, born in Lausanne, has, for instance, constructed a non-linear theory of massive graviton that could describe the accelerated expansion of the universe as a gravitational effect. This theory would eliminate the need for dark energy. If an experiment can, one day, decide which theory makes the best prediction, an existent object could suddenly become non-existent.

For a non-physicist, this extended definition can confuse because it relies on previous theories to assert the existence of objects, "out there." What if our mathematical models change? Are we not creating a risk of self-referral where things exist because our theories predict they exist? (6.10).

The objective existence of elements in the universe would depend on our capacity to use those unobserved entities to justify observed events. It is a curious mix-up of map and territory.

Brain-centrism supports extended existence. It asserts that any narrative is about our representations and it should make us very careful with non-observable objects. This is true for objects that are equivalent to their verbal definition as quantum objects. Their definition or the mathematics that predict their behavior acts as expectations for our mental space. No sensory information will contradict those expectations, as the object is not visible or measurable. We then take the risk in ERP to attribute to the universe a property that belongs to our brain. (4.12)

* *

6.4 Causality and special relativity

In Special Relativity (SR), the passage of time depends on the speed of the observer. Closer he gets to the speed of light, slower will be his watch when viewed by an external observer. The

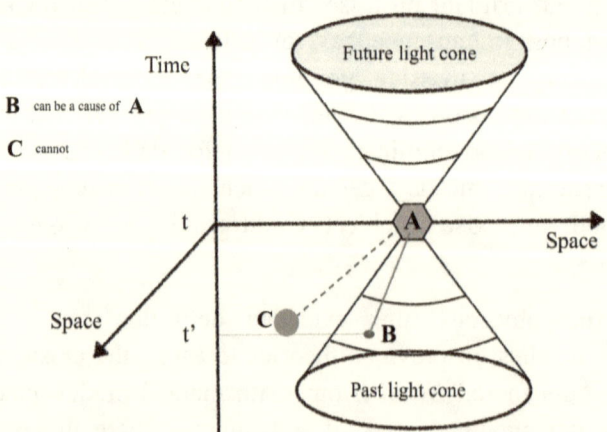

Figure 6.4.1 : *Einstein's light cone*

Langevin paradox illustrates this effect. A man taking a one-year trip into space with his fast spaceship will notice, when he comes back one year later, that all humans on earth have aged 50 years.

But Special Relativity, although it neglected gravitational effects by hypothesizing non-accelerating frameworks, already caused deep changes in our understanding of reality.

Notions like time and space changed radically and became relative, we excluded ether. The "simultaneity" of two events became a notion dependent on the respective speed of the observers. The notion of physical causality had to be revised.

To every event A at time t, one can associate a cone with its summit on this event (6.4.1). Suppose B is an element at time t' in this cone. A photon coming from B could influence A. Indeed, it would have had the time to travel from B to A. Any element outside the light cone could have no influence on A. Causes for A are thus all in the light cone having A as the summit. Samewise, A can only influence the events in the future part of the cone. The light cone represents observable events. The region out of the light cone contains non-observable but hypothetical events, useful to explain observable events. It might well be that C is in the light cone of B, for instance.

For instance, the reason physicists distinguish "the universe" from the "observable universe," the part of the universe that is inside our light cone.

In (3.9), we have shown that causality is a metal space function; we will further confirm it.

In SR. an event X can happen before an event Y for observer A and after Y for an observer, B. This "reversed order" in their perspective depends on their respective speeds.

Here, for A, X could cause Y, while for B, Y could cause X.

But SR. is a third-party ERP perspective of an observer C. For C causality (where the cause precedes the effect), disappears. So does simultaneity.

However, for A and B, in their first-party perspective, causality remains always respected whatever their speed.

An observer by himself will never experiment with time distortion, he will never see his watch running slower or faster. To notice that their watches don't show at the same time, A and B will have to communicate.

SR. is a model build from the perspective of a third-party C. This third party can never exist at the moment of the observation. What makes SR. difficult to understand is that we describe in C's third-party perspective, and at the same time we want to understand it from the first party perspective of one observer.

A and B cannot agree on the time separation between two events, neither can they agree on the distance separating two objects. This shows that time and space are not "usual" measurable physical entities. In a third-person perspective, they must be a function of the observer's speed, as the equations of SR. describe them.

However, in the first-person perspective, time and space do not depend on speed. The difficulty appears when two observers communicate. When A and B communicate, they need a common model that makes sense of both A and B's observations. That is the genius intuition of Albert Einstein's construction. Einstein builds a relative time and space, never directly observable, but useful to explain what A and B observe when they compare their watches. We end up having two times and two spaces. The first party ones, who are measured during the observation, and the third party ones, whose "existence" relies on the extended definition of existence. (6.3)

In the Langevin paradox, both the traveler and the earth residents have seen the time pass by normally, nothing changes for them. When they meet, they notice the traveler did not age as much as the residents. His watch shows that one year has passed by and the resident's watches show that 50 years have passed by. All was normal for each of them until they meet. The surprise when they meet happens because they are convinced that there is an absolute time equal for everybody. What makes it paradoxical is that they now realize that this absolute time cannot be the one they have experienced!

The discovery of the constancy of the speed of light and Einstein's special relativity has had a tremendous impact. It obliged us to change our worldview on fundamental conceptual pillars such as causality, space, time, and the relation between matter and energy. "Out there" did not change. What changed because of this additional information is our knowledge of "out there." The discovery of some information inaccessible in our day-to-day life obliged us to interpret common things differently.

We never observe the past, "out there." Nor can we observe the future. We only observe, "nows." Observing is always a first-person activity of the natural world. (Figure 5.3.2)

When you observe a distant galaxy, what you see is happening now. A third-party perspective (a verbal world perspective) could infer that, as the light took a billion years to reach us, what you observe now happened there a billion years ago. However, that is a mental inference, not an observation. The first and third parties claim different things. The first party is directly experiencing a "now" observation. The third party is inferring from a verbal world theory when (in our time) was encoded by the information that this light carries.

What we observe now is always happening now (the territory). Theories may infer from this observation something about the past (a map). It is a simple abuse of language to say we are observing the past when we are inferring it theoretically.

Any knowledge by acquaintance happens here and now. On the contrary, knowledge by description can be about the past or the future and anywhere, even outside of the universe. Past and future cannot be part of first-party experiences. They are only third-party conclusions useful to explain observations.

We have become so used to our "verbal world," we confuse it with "out there." For some of us, it has become difficult to see the difference between a mental inference and an observation. This confusion is very common even among scientists. They trust their science so much that, for them, it is "out there."

ERP sounds so logical along as we don't realize that we are confusing map and territory. We treat theoretical inferences as if they were observations.

To avoid confusion, it is useful to train our "consciousness of abstracting." This must be a "being line" training by meditation or acquisition. Descriptions will not help.

Let's consider again the figure (4.12.1) ERP explanatory landscape is between arrow 1 and arrow 2 on the right side. Brain-centrism suggests extending it to the left side of the picture. Limitations to knowledge car happen by characteristics on both sides. For instance, mental blind spots (on the left side) will have to be explained by ERP by an event "out there" on the right side. ERP might attribute to the universe characteristics of the observer.

The brain-centric view makes clear the difference between first-party and third-party views. It also clarifies causality as a property of the verbal world rather than out there.

* *

6.5 Understanding understanding

In a famous citation, Einstein declares:

Any fool can know. The point is to understand.

If one can test knowledge, testing understanding is much more complicated. Understanding is a feeling. We can get it from the first person's perspective, but it's difficult to measure as a third person. Like any feeling, it can take different forms and is not definable by description. It can refer to a narrative, but also a range of non-verbal sensory signals or feelings perceived by the mental space. It's self-referral, one understands,§ one understands.

By reading a book a second time after a few years, one realizes that the first time one had not understood it the same way.

John Kunios is a professor at Drexel University, he declared in an article in Aeon:

"The fact that evolution has linked the generation of new ideas and perspectives to the human brain's reward system may explain the proliferation of creativity and the advancement of science and culture."

He could have added that understanding is also rewarded by the brain. Probably because it's a survival asset. In the verbal world, we increase our understanding by building causal chains and relating the subject to known facts. In the natural world, the feeling of understanding comes through analogies, often visual. Understanding plays the role of a fill-in reducing brain entropy and our increasing our prediction capacity. Explanations or analogies can trigger understanding but are not the understanding itself.

Understanding is not algorithmic, there is no "how to." Having been in teaching, "how do we understand?" has remained an important question for me, like for all teachers. We have ways to facilitate the understanding by examples and images of the subject, but not of the understanding itself. Because itself as an "objective" third-party entity does not exist.

I remember Roger Penrose at a conference in Zürich; he was explaining to an audience of unconvinced computer scientists that understanding was non-computable. He was correct, and his demonstration was clear: understanding is not algorithmic. ERP physicalists in the audience refuted Penrose's argumentation, based on Gödel's incompleteness. Computer scientists are so used to define feelings with descriptions. They seem to have lost the basic understanding of a feeling. That is terrifying when you imagine how much the algorithms these people produce now control our lives, just as terrifying for me as I imagine the inquisition was in the Middle Ages for a non-believer.

One cannot know if someone else has understood or how he has understood; we have to ask him. Like in an audition test, the doctor asks you to raise your hand when you hear the sound. He has no way to measure our experience by himself. A person can evaluate his understanding from a first-party perspective. From the outside, it is not measurable.

In the natural world, a series of connections to more in-depth Gödelian information represents understanding. The feeling appears when one relates an unknown event/representation to a previously understood phenomenon. Understanding unites, consolidates, and lightens up our mental space, preparing it for new questioning. Understanding triggers a feeling of conviction and knowledge.

Any intensional definition of understanding would not only be a terrible simplification but would also create antinomy because of the recursiveness of the word.

So why are super smart computer scientists, like Schmidhuber, claiming that they can program feelings? They add critical words if the feeling is properly defined. That is the danger (5.14) of the "imitation fallacy," replacing a feeling with a brief series of words, then pretending that this series of words represent the feeling.

Convinced computer scientists have another way to put it. They declare, *show me something a brain can do and a computer cannot*. They express that way their faith in ERP. What "exists" is what we can show. But remember what we said in (2.8) on materialization. Whatever we build has to go through digitalization and a blueprint process in the verbal world, before we encode it into the matter. Whatever visible object or event we produce "out there," a computer can also do. But it cannot develop emotions and or feelings, love his children, understand, or prove mathematical conjectures, or play with infinity, etc. But we cannot show all these "things," although they are the most important things in humans' lives. They belong to the natural world, and computers don't have it.

Many scientists have made a habit of building up thought experiments by imagining a situation and trying to figure out its consequences. Einstein explained that at 15, he imagined himself sitting on a ray of light chasing another beam of light and wondered what he would see. Some thought experiments have later found a realizable version. The method is fruitful and very illustrative of how imagination can trigger understanding.

Sometimes the precise point we do not understand is unclear. If we can imagine a thought experiment, it may underline what piece is missing or what unconscious hypothesis we have made that should not be there. That is why often scientists love good science fiction. By accepting some dose of fiction, it can examine how science works in that situation. Sit on a light ray to realize that light cannot stop.

That justifies Einstein's observation we stated to start this paragraph. Knowing and understanding are, in consequence, different things. In classrooms, students can know (repeat) even when they have not understood. We will then not anchor this knowledge with feelings; although they can repeat it, it will disappear. Analogies, pictures, examples, experiencing are useful tools for the teacher to trigger understanding. Verbal explanations are not sufficient, especially for younger children, acquisition is critical.

Create by oneself an explanation is the best way to generate understanding. Explaining to others and confronting and convincing them is the best possible way to deepen our understanding.

A good teacher would take advantage of the excitement and satisfaction generated by creative activities, he would use the

Socratic method and ask a question and let the mental spaces of his students enjoy searching, creating, and finding. A good Sherlock Holmes detective or a scientist would need more than an initial analogy. He wants the piece of the puzzle to fit; however, you turn the game and from whatever angle you look at it.

* *

6.6 Intuition

We know our mental space feeds at two sources of information. (3.5) Even with no sensory input, it continues to build representations based on memorized information and emotions. That is how imagination works. Since Kuhn and the end of logical positivism, we now accepted that creativity and intuition have their role in producing science.

Rules and methods do not exist for every situation. Uncomputability (4.8) shows that even in mathematics, situations exist where no laws are possible (4.7). No logical reasoning can then suggest a solution, and we must rely on our moral sense, on our "intuition," or "good sense" to decide. The verbal world is confronted with a blind spot. (6.10) The mental space must rely on the natural world. (4.2)

For every representation (R), our mental space builds up a *logical context representation network,* Net(R) associated with (R). (1.7) We know that Net(R) extends to both the verbal and the natural world. We also know that Net(R) depends on the context and the general mental state. That explains why an intuition or a creative idea will turn up at a precise moment.

What we call higher brain functions such as creativity and intuition to appear because Net(R) associates to (R) a network that had previously not been formed or associated. This happens at a precise moment when the mental state and its environment present a suitable "alignment." Brain regions will synchronize that are not usually working together.

Net(R) can be imagined as a "search engine." However, its search is not based on researching an existing database for words or concepts. Net(R) will build analog and causal associations, recursively guided by the general state of the mental space and the sensory inputs. It can thus search for unknown things. Things that do not pre-exist in a database.

Net(R), as we often noted, covers both worlds. However, intuitive and creative ideas appear through analog associations of the natural world. Like with Kuhn's context of discovery.

Net(R) is a non-reversible procedure. It is impossible to reproduce the state of the mental space and its environment that triggered the intuition. This irreversibility assures that the procedure is non-mechanical and uncomputable. No materialized machine can produce it, including computers. (2.8)

The brain is a dissipative system, it uses incoming energy to self-organize, create new connections, and lower its entropy. Part of the inflowing energy is converted through self-organization into Gödelian information. (2.11) This novel information is physical and materialized in the brain by the connections producing the Net(R)'s. Our "non-reversibility" is what ensures our creativity and our intuition.

But it also ensures that no reversible machine could simulate the brain. Irreversible machines would be useless.

Verifying proof can be done algorithmically step-by-step. But the proof was not conceived that way. It was dreamed, it came as a long succession of "revelations," of sudden inner convictions, of visions, of "Eurekas." The creative process was mainly emotional, visual, tactile, and Gödelian, it needed Net(R) to plunge into the natural world.

Brain-centrism suggests that education and research should consider this asymmetry and could largely profit from neuroscience's findings on how the brain builds Net(R)s.

Srinivasa Ramanujan, one of the most astonishing mathematical geniuses of all times, claimed that he had received in his dreams a vision of entire rolls of complex mathematical formulas given to him by the Goddess Mahalakshmi. This "explanation" of the source of his immense talent was the third source we have evoked.

Often intuitions are so surprising for the discoverer himself, he cannot believe that his brain did it. Intuitions are by definition not algorithmic, like gods they put us in touch with something that belonged to an invisible world.

Henri Poincaré had already similar ideas at the end of the 19th century, although not much yet was known about the brain and the mental space:

" ... for that they needed a guide. This guide is, first, analogy. For example, one of the methods of demonstration dear to analysts is that founded on the employment of dominant functions. We know it has already served to solve a multitude of problems; in what consists then the role of the inventor who wishes to apply it to a new problem? Initially he must recognize the analogy of this question with those which have already been solved by this method; then he must perceive in what way this new question differs from the others, and thence deduce the modifications necessary to apply to the method.

But how does one perceive these analogies and these differences? In the example just cited they are almost always evident, but I could have found other where they would have been much more deeply hidden; often an uncommon penetration is necessary for their discovery. The analysts, not to let these hidden analogies escape them, that is, to be inventors, must, without the aid of the senses and imagination, have a direct sense of what constitutes the unity of a piece of reasoning, of what makes, so to speak, its soul and inmost life."

Poincaré explains the role of analogy and how a mental preparation at the verbal level can favor the construction of useful Net(R) able to perceive differences.

But Poincaré also says: "*[one must] have a direct sense of what constitutes the unity of a piece of reasoning.*" That shows that analogies in play in intuition cover multiple dimensions. The similarity can, for instance, concern the content, the structure, the environment, the procedures, the notations, or even some emotional reaction when facing the question. That explains the importance of the state of mind and the context. If a simple detail can trigger the analogy, the experience will allow us to spot it out, give it value and attention and extend it to new regions of the mental landscape. A proper internal state of the mental space is critical to trigger its tendency to draw analogies. We must feel free from heavy and rigorous rules, like when a child is ready to play or ready to make jokes. Reaching this kind of state needs us not to be under any kind of pressure. Often education for practical and economic reasons concentrates only on "*How to…*" There is no how-to for creativity and intuition.

"*A direct sense of what constitutes the unity*" refers to Gödelian information, Poincaré uses "direct" and "sense" in place of "symbolic" and "knowledge." Intuition is not about reasoning, but about triggering direct sense, making improbable connections between unconnected things, seeing unity between things seen as different, synchronizing to create unity.

If intuition or moral sense can often lead to understanding, it can also sometimes be misleading. This happens because it is using brain patterns and representations inappropriate to the situation under examination. It's because of some lack of knowledge of the subject. The representations used by intuition are sometimes outdated or miss some critical information. That explains why, to get our brain to produce valid intuition, one needs a lot of preparation and extensive studies. How else could we know what makes up the "*unity of a piece of reasoning*"?

Most of the technologies we use today would appear impossible because contrary to the intuitions of humans, only a couple of dozen years ago. Most of the present scientific thinking has now diverged from ancestral intuitions. Without a deep-rooted culture in a specific subject and its environment, most (but not all) of our intuitions will be useless.

Poincaré suggests the following mathematical example:

"We know there exist continuous functions lacking derivatives. Nothing is more shocking to intuition than this proposition which is imposed upon us by logic. Our fathers would not have failed to say: 'It is evident that every continuous function has a derivative, since every curve has a tangent.'

How can intuition deceive us on this point? It is because when we seek to imagine a curve, we cannot represent it to ourselves without width; just so, when we represent to ourselves a straight line, we see it under as a rectilinear band of a certain breadth. We well know these lines have no width; we try to imagine them narrower and narrower and thus to approach the limit; so we do in a certain measure, but we shall never attain this limit. And then it is clear we can always picture these two narrow bands, one straight, one curved, in a position such that they encroach slightly one upon the other without crossing. We shall thus be led unless warned by a rigorous analysis, to conclude that a curve always has a tangent."

In mathematics, we define objects intensionally. (6.5) To understand the specified object, one needs to attach an analog representation, often a visual representation and feelings. Sensory perception alone does not guide us for abstract non-physical objects. Proper analog representations will generate pertinent intuitions; improper ones will produce misleading intuitions.

One can have developed proper representations that give us a good understanding of hydrodynamics. Then try to apply them to electromagnetism, where they will provide us with a limited understanding and wrong intuitions.

Poincaré continues:

"Pure analysis puts at our disposal a multitude of procedures whose infallibility it guaranteed; it opens to us a thousand

different ways on which we can embark in all confidence; we are assured of meeting there no obstacles; but of all these ways, which will lead us most promptly to our goal? Who shall tell us which to choose? We need a faculty which makes us see the end from afar, and intuition is this faculty."

This is a compelling remark. Even if a problem accepts an algorithmic solution, if a "step-by-step method" exists, intuition is, however, necessary to spot out in advance what method will lead us to the result. A mathematical proof can always be verified algorithmically. Because of the Halting problem, we cannot find it algorithmically. Explaining the past is easy, predicting the future is more complicated.

Intuition is a central capacity of the mental space that a digital computer cannot have, lacking a natural world.

Intuition, when viewed from a third-party perspective, seems to be an unjustified "guess." Evolution has given us the capacity not to get stuck by equivalents of the Halting problem in the verbal world. The brain escapes the blind spots of the verbal world by referring to the natural world. The natural world works then as a Turing Oracle.

This Oracle makes a guess or rather generates an intuition suggesting a way to go when nothing algorithmic can guide our steps.

This approach makes sense in the brain-centric perspective. However, we must learn how to use this hybrid analog-digital oracle. For it to be interesting, a solution must be compatible with the complete structure; it can otherwise be misleading. That is why valid intuitions only manifest only one has accumulated sufficient knowledge. Studies show that we rarely resort to rational thinking in day-to-day life. Most decisions are dictated by patterns, like stereotypes, or are based on feelings or intuitions.

Turing himself has insisted that "intuition" is present in every part of a mathematician's thinking. For instance, when a mathematical proof is formalized, intuition has an explicit role in the choice of successive steps. However, Turing did not say what the brain was doing in a moment of such "intuition." In his 1939 paper, he distinguishes "ingenuity" from "intuition." Ingenuity is what a clever Turing machine can have. Intuition, he says, is "something else":

"Mathematical reasoning may be regarded ... as the exercise of a combination of ... intuition and ingenuity... In pre-Gödel times, it was thought by some that all the intuitive judgments of mathematics could be replaced by a finite number of ... rules. The necessity for intuition would then be entirely eliminated. In our discussions, however, we have gone to the opposite extreme and eliminated not intuition but ingenuity, and this in spite of the fact that our aim has been in much the same direction."

In 1939, Ludwig Wittgenstein conducted, at Cambridge, a set of lectures on the foundations of mathematics. Alan Turing took part in some of these lectures. The discussions between Wittgenstein and Turing have fortunately been recorded. They center on the notions of discovery and invention in mathematics. Turing still seems sometimes to believe that discovery is algorithmic. Wittgenstein insists that even if "a posteriori" mathematical operations can be presented algorithmically, "a priori" needs non-algorithmic creativity.

Albert Einstein is known for having declared that *intuition is the highest level of intelligence*. The precise quote goes much deeper and is a profound critic of our society:

"The intuitive mind is a sacred gift and the rational mind is a faithful servant. We have created a society that honors the servant and has forgotten the gift."

Einstein says that, when confronted with a problem he could not solve, he had developed the habit of taking a nap. Sometimes he would then wake up with a solution.

Bernhard Riemann's work was literally based on intuition. He had developed this type of reasoning during his discussions with

Dirichlet. The substance of his ideas was not hidden in lengthy computations and his writing fell below the rigor usually required from mathematicians. Moritz Stern (1807–1894), a student of Gauss in Göttingen, was perhaps the first to realize that Riemann was a genius. He later described the young Riemann by saying that he:

"... Already sang like a canary."

Srinivasa Ramanujan, as we know, was a profound Hinduism believer. He famously declared:

"A mathematical equation makes no sense for me unless it represents a thought of God."

Ramanujan did not take much care to prove his results (more than 3600 theorems and complex formulas), he knew they were correct. It took some 60 years after his death, at the age of 32, to finish proving formally all his proposals.

August Kekulé discovered the ring structure of the benzene molecule. He attributes his discovery to a dream he had in which a snake was eating its tail. When he woke up, he realized that the dream was indicating the circular shape of the benzene molecule.

An article that appeared in Quanta magazine describes how the field medal mathematician Terence Tao got an idea to (partially) solve the "3x+1" conjecture that is known as the *simplest unsolved problem in mathematics*. The conjecture posed in the 1930s and also known as Collatz conjecture is so simple that anyone can try and prove it. It states:

Start by choosing any positive integer. If it's odd, multiply it by 3 and add 1. If it's even, divide it by 2. Starting with the number obtained repeat the procedure. The conjecture asserts that you will eventually reach number one. Tao, using his trained and informed intuition, saw an analogy between this conjecture and methods to solve partial differential equations. That leads him to his partial solution.

Henri Poincaré explains that after a long period of hard and fruitless effort to solve a problem, he took a break. As he was stepping on a bus, he made one of his most important discoveries. The solution came to him out of nowhere. Poincaré describes how this solution appeared, along with the feeling of certainty that it was the correct approach!

On another occasion he explains:

"For 15 days, I strove to prove that there could not be any [mathematical] functions like those I have since called Fuchsian functions. I was then very ignorant; every day I seated myself at my work table, stayed an hour or two, tried a great number of combinations and reached no results. One evening, contrary to my custom, I drank black coffee and could not sleep. Ideas rose in crowds; I felt them collide until pairs interlocked, so to speak, making a stable combination. By the next morning..."

Paulo Coelho, the famous Brazilian author, describes in his book "*Maktub*" how Roger Penrose had one of his brightest intuitions. He was walking and discussing with friends. The conversation stopped while they were crossing a busy road:

"*I remember, declared Penrose, that while crossing, a fantastic idea came to my mind. But arriving on the other side, the discussion continued and I lost the idea a just had some seconds before.*"

At the end of the afternoon Penrose was feeling a bit euphorically, without understanding why:

"*I had the sensation that something important had been revealed to me.*" He reviewed every minute of what had happened that day. And when he remembered crossing that road, the idea came back to his memory and he wrote it down. It was a revolutionary idea on black holes.

Let us cite the Field medalist Mathematician René Thom:

Whatever is rigorous is meaningless.

* *

6.7 Free will

"It is a psychological fact that we believe we have the ability to control and modify our actions by the exercise of 'will,' and in practical life all sane men will assume they have this ability," declared Sir John Eccles (1903–1997). John Eccles was a very famous Nobel prize winner, Australian neurophysiologist and philosopher.

"Freewill," represents a feeling and a belief. We know feelings are not definable by description. As a feeling and is generated in Gödelian information in the natural world. It is not observable. However, we consider it the basis for all our freedoms. Depriving someone of liberty is the most obvious punishment. We consider freedom as one of their most valuable attributes. The ideas of choice and responsibility are the roots of our economic and legal systems. We base democracy on the capacity of making a choice.

Freewill was at the core of Descartes's mind and body problem and the major subject of Schopenhauer's preoccupations. It requires the capacity of mental causation. In (3.11), we described how thoughts and feelings can act on the body and thus on "out there."

Most humans feel the free will experience as "true," and build their lives on that belief. They feel they can choose and decide. They recognize they have to use the laws of physics and cannot go against them. But for them, the constraints imposed by these laws give them space to move within a framework.

However, this is not what many physicists believe.

The ERP perspective estimates that a Laplacian determinism (5.11) cascades down the causation chains. The conservation laws leave no place for effects with no causes. Everything we decide and do must have a cause. ERP rejects mental causation (see figure (3.11.1)) the cause must be a material cause. ERP would then look inside the brain and search for the previous step of the causal chain that led to the decision.

ERP materialists will consider that free will, like consciousness, is an artifact, an illusion associated with neural activity. They explain we experience this illusion because of the restricted perspective of first-party view and oppose to it the lack of measurability of this feeling.

The famous philosopher Daniel Dennett (1942-) is a compatibilist, he believes, as did Henri Poincaré that free will and determinism are not incompatible, as he explains in his book, "Brainstorms."

Many thinkers have tried to reconcile the feeling and the third-party antinomic definition of free will by referring to quantum indeterminacy, hoping to find there a Gap to give a place to free will. Quantum indeterminacy is itself a mental abstraction of the verbal world.

For most of us, that one cannot choose between tea or coffee because we live in a deterministic world seems ridiculous and illustrates the antinomy between a feeling and a third-party ERP description.

The Geneva quantum physicist Nicolas Gisin, who has experimentally confirmed Bell's inequality, declared:

"I know that I enjoy free will much more than I know anything about physics. Hence, physics will never be able to convince me that free will is an illusion. Quite the contrary, any physical hypothesis incompatible with free will is falsified by the most profound experience I have about free will."

Belief in the Laplacian universe will never change our reaction when facing a lion. Laws are perhaps deterministic, but we still cannot predict what will happen. So the future in the verbal world of Laplace is perhaps determined. However, for our first-party view, better trust what millions of years of evolution have taught us: escape.

* *

6.8 Time

Most animals immediately react to movements and changes in their environment. Our Savanna human was permanently on alert, tracking with all his senses for any changes that could mean danger. (1.1). Nature has captured the importance of movement detection by dividing the retinal ganglion cells into two general categories: some cells with higher bandwidth are specialized in detecting movement.

Expecting changes, memorizing and continuously updating our representations, have translated over the millennia into a feeling. Memory separated past from future, body constraints, circadian rhythms, and repetitive natural events, modulated this feeling who became best described by the words "time flow." Our explicit memory allows us to remember contextual information and associate internal sensations with external events. The quasi-permanent activation of the time flow feeling reinforced its connections to most representations. Memories organizing themselves according to time flow decreased the brain's entropy, transforming it into a fantastic survival tool. This timely organization, by improving the quality of our expectations, has been critical to our adaptation capacity and had shaped our general worldview. Most evolved mammals have developed time flow organization capacities, although certain aspects can differ greatly from one species to another. This very much depends on how the memory works. Mammals mainly use implicit memory (2.1) and learn only by imitation and acquaintance. They remember things by associating smells, sounds, and activities with specific representations.

After an extensive study in the Serengeti Park, Fritz Walter, who was studying how gazelles estimated the security flight distance, wrote:

The flight distance varies with the different predators according to their danger for gazelles. Experiments showed that adult males in herds (bachelors) have a lesser flight distance than females in herds, ... Thus age, sex, and social status are all significant factors influencing flight distance.

Mammals have learned long ago how to estimate speed. But interestingly, most mammals have not fragmented their reality in the same distinct conceptual categories as we humans have. Flight distance for a gazelle will depend on its escape speed or the predator speed, but it will also relate to the social status or the age of the animal, and the wind direction, or the nature of the ground.

Imagining the mental space of our Savanna man is difficult. Similar individuations did not fragment his "reality" as ours. His vision was much more holistic. Symbolic language reinforced explicit memory and the fragmentation of our worldview. Abstract concepts, including time, became entities for their own sake, separated from movement.

The abstract concept of time, although it remains related to the feeling of time flow of the natural world has now its verbal world existence. Isolating time, as if it was "objective" and seeable, did not eliminate our ancestral feeling of time flow.

We progressively learned to live with the antinomy of two times: our ancestral feeling of time flow and common consensual verbal world abstraction of time, the time of the clocks. Time would become the first feeling we have materialized with "clocks," to then impose ourselves to structure our lives according to this materialization.

How did we measure a non-seeable abstract entity? We never measure it. We compare various low-level representations, such as unique positions of the sun in the sky or of rotating needles. We associate these ordered representations with other events and use these representations as a reference base. "The sun is getting lower on the horizon, winter is coming." Through these associations, we materialize time flow, we have projected what is a feeling onto visible events.

We find the earliest sundials known date some 1500 BC in ancient Egyptian and Babylonian astronomy. Mechanical clocks only appeared three thousand years later in the Middle Ages. The crucial aspect is the fact that we have associated a number with what was a feeling. We have transformed time into a common digital currency ruling our societies.

As feelings cannot be intensional defined, science (and society) adopted this number as a definition. However, it does not capture what we feel as time flow, creating an antinomy.

What we experience is not always what the clock says! As for mammal's flight distance, our feeling of time flow is also related to a variety of internal mental events: are we enjoying the moment, are we expecting something?

We are now embedded in this "objective" digitalized time that we are convinced it is a "natural" attribute of "out there," although nobody has observed it, as it is a mental construct. Our ancestors before language lived the "now," as the famous philosopher and author Alan Watts (1915–1973) wrote:

"We are living in a culture entirely hypnotized by the illusion of time, in which the so-called present moment is felt as nothing but an infinitesimal hairline between a causative past and an absorbingly important future. We have no present."

*

ERP considers time as an "existing" physical entity. As we cannot observe it, this "existence" can only be of the extended type and must belong to the Platonic realm associated with physicalism. (5.1.)

Einstein defines a time in Special Relativity as what clocks measure, to have a usable, numerical ERP definition. However, this definition is circular, as any definition of primitive concepts. Time is defined with clocks and clocks are themselves defined as instruments built to measure time.

But this circularity should not astonish us. Physicists need digital units, and the definition of these units is always circular, as it needs to connect a mental concept with some observable event "out there." In 1799, the meter was defined in terms of a prototype meter bar deposited in Paris. It is now defined as the length of the path traveled by light in a vacuum in 1/299,792,458 of a second which is also a circular definition. The circularity of these definitions only expresses the impossibility to bridge what is mental with what is out there without a measurement. Measurement is the only connection between map and territory.

Other ERP physicists have different views on time. The British physicist and author Julian Barbour (1937-) described his view of time in a 1999 famous book: *The end of time*. He suggests that there be no time "out there." His thinking is in the Machian tradition; Ernest Mach rejected the existence of entities that are not directly observable as being "non-physical." However, Barbour's position remains very controversial. Sean Carroll in his book *The Reality of Time*, in 2015, says: *I think that these folks are working hard to tackle a problem that has already been solved.*

Leibniz had an original idea of space and time. He rejected the Newtonian *"stage"* on which everything happens. He would not have agreed with Einstein's *"what clocks measure,"* nor with Carroll's. His idea much better captures our feeling of time flow and fits with brain-centrism.

He writes in his third paper:

As for my own opinion, I have said more than once, that I hold space to be something merely relative, as time is, that I hold it to be an order of coexistence, as time is an order of successions.

* *

6.9 Limits to scientific knowledge

What are the limits to what we can know? Sir Martin Rees, the British Royal Astronomer, famously declared:

"There is no reason to believe that our brains are matched to understanding every level of reality."

Science has made remarkable progress, and we seem to close certain knowledge gaps. However, we should remember that every answer makes us aware of many additional questions. We know much more than we knew 500 years ago, but we have more open questions we could not even plan. Knowledge is not like a book with a beginning and an end. For each page you read, a hundred new pages add up.

Brain-centrism and ERP may have different explanations for the origins of our limits to scientific knowledge.

One could distinguish four types of limits to scientific knowledge.

1- Limits because of the scientific method itself.

2- Limits because of entropy and lack of access to information.

3- Limits because of the fabric of reality out there.

4- Limits because of the structure of our mental space.

1- *Limits because of the scientific method.* Science does not cover all possible knowledge. It is seeking regularities to extract predictive laws. Its domain is the limited zone of predictable events. Un-measurable events are not part of scientific knowledge. That excludes de facto experienced events. ERP science that only considers measurable causes will not embrace the source of its existence. Brain-centrism considers this first type of limitation as to the requirements we have chosen for some knowledge to qualify as scientific. These requirements have changed in the past and will continue to change. If science ignores an event, it does not mean it has no existence.

2- *Limits because of entropy and lack of access to information*. For various reasons, we cannot collect information. For instance, because of the dispersion of information through entropy and entanglements, or because of Shannon's noisy channel limitations (3.4), or because of the difficulty to access information without altering the event, we are observing.

We rely on instruments to increase our capacity to access information. However, in physics, deeper, we want to dig into the structure of matter, higher are the energies involved. The total CERN energy consumption is 1.3 TWh per year, the one of the LHC is equivalent to 600 GWh per year. Geneva canton uses about 3 TWh per year, with 500 thousand inhabitants. In neuroscience, new scanning instruments have allowed a better evaluation of what is happening inside the living brain; their resolution remains still very weak. The typical fMRI spatial resolution is pixels of 3 mm to 4 mm. One cubic millimeter of the cerebral cortex contains roughly 50,000 neurons, each of which establishes approximately 6,000 synapses with neighboring cells. We are very far from being able to observe in real time what happens in a single brain cell! Furthermore, the temporal resolution is terrible.

A critical point in the measurement/experimentation process is that the same party, the mental space, is individuating the elements, conceiving the experiment and the set-up, doing the measurement, and analyzing the results. No true third party is there to evaluate the operation. As David Bohm had remarked, the risks of confirmation biases and self-referrals are high.

The scientific method requires the experiments to be repeated by distinct groups to eliminate biases. Thousands of physicists sign large experiments requiring huge instruments, like CERN's Large Hadron's Collider. How can the traditional peer review process then function and what distinct group will effectively reproduce the experiment?

3- *Limits because of the fabric of reality out there*. In ERP these limits appear for several reasons. Because of the limited speed of light, the notion of simultaneity can disappear. Because entropy or entanglement increases like under our first point, it is impossible to collect all the information on a physical object. Heisenberg's uncertainty expressed most of the limitations in this category. It concerns the quantum level. In the same category are Bohr's complementarity principle and the no-go theorems such as Bell's theorem that we will examine in part two of Brain-centric. These limitations are not because of the structure of reality out there, but of the mental space itself because of its blindspots. (6.10)

4- *Limits because of the structure of our mental space.* ERP does not formally recognize such limitations as they limit the investigations to the right side of the figure (4.12.1). For brain-centrism most limits involve mental operations such as individuations, expectations, first-party view, fragmentations, abstractions, language, blindspots, …

The study of these brain-related limitations will become critical for future knowledge.

* *

6.10 Self-reference and blind spots

Self-reference is, as its name shows, a property that only mental verbal abstractions can have. "Out there," there is no self-reference because "things" don't "refer to" other things. Only representations "refer to." Some self-references apply to an entity without referring themselves to this entity such as in the proposition: all matter is made of atoms, it applies to this book without explicitly referring to it. We will say that these propositions self-apply.

The Greeks already knew that self-reference is a linguistic phenomenon that causes paradoxes. Epimenides' liar paradox: *all Cretans are liars*; its formulation by Eubulides[66]: *a liar claiming that he is lying*; the Socratic paradox: *Socrates knows*

---

[66] https://en.wikipedia.org/wiki/Eubulides

he knows; are variations of similar self-referential expressions. Russell's paradox, the one he expressed in his letter to Frege (4.5) uses a self-apply paradox. Gödel based the original proof of his first incompleteness theorem on the self-apply sentence: *This sentence is not provable.* (4.5).

Ancient Greek Skeptics like Xenophanes and Democritus had raised the infinite regress problem. They explained that to know something (A), it is not enough to believe (A), one must have a good reason (B(A)). The good reason must be something else than (A), it is "about" (A). But to know (B(A)), one must have a good reason (C (B(A))), different from (B(A))., etc. The skeptics argued that no belief is known unless an infinitely long chain of other beliefs supports it. They claimed it is not humanly possible. Therefore, we know nothing!

The infinite regress of causes uses self-reference or self-apply. It is not generated merely by the assumption that every event must have a prior cause, but by the assumption that every event must have *a distinct* prior cause. Remember that causality is a mental operation that allows the mental space to organize itself (3.9) and that infinite regress is because of our capacity of abstracting indefinitely.

In (3.7), we observed that reduction and inclusion lead to problematic infinite regresses. ERP has to block regresses somewhere as it considers them referring to "out there."

Democritus blocked infinite regress of the self-apply sentence, *what is it made of*, by considering atoms? Plato did it with his platonic elements and the standard model with quarks and electrons. A regress is blocked by asserting that there is at least one occurrence for which the property of having a distinct prior origin or cause is not self-applicable. The blocking element is then "*made from nothing*" but itself.

We can observe that a blocking element cannot refer to something directly observable. If it were observable, one could continue asking "what is it made of" and pursue the regression a step further. The origin of such chains is thus always a mental entity that refers to a non-observable entity "out there" that has no cause. It is not the mental space that causes the blocking element, but ERP constraints that require every step of the abstraction chain to represent something, "out there."

The universe is an example of a blocking element for the self apply sentence, what is it included in? We already know how paradoxical this concept of "the whole" can be. The universe is non-observable, as it has no "outside" to observe it from. Don't forget we are in the verbal world where descriptions are expressed in a third-party perspective. However, the limited speed of light and Einstein's light cones (6.4) have allowed us to define the *"observable universe"* that is a blocking limit. The observable universe is thus a non-observable mental abstraction[67].

The self-apply sentence, *who made it*, assumes that things are always made or created. ERP requires a blocking concept for this sentence. Our mental space has generated the fill-in concept of gods. Gods must thus either not have been made and be eternal, or have made themselves[68].

Physicists like Lawrence Krauss, who declares himself an atheist, wrote in 2012 a book entitled A Universe From Nothing. He shows how a universe can emerge out of "nothing." For physicists "nothingness" is the lowest-energy state of empty space or the ground state of the universe. This is an ERP third-party observation, where the observer is out of the universe full of nothing.

Descartes' I think therefore I am seemed to him, and to most of us the very source of the undoubtedly. He based his discourse on this simple observation. However, this sentence is a self apply

---

[67] The word observable is used here with two different meanings. In the "observable universe" it's an observation from the inside. In "non-observable" it's a third party observation from the outside.

[68] Remember we are speaking here of the limits of knowledge in the verbal world under ERP. Many people have become atheist without noticing that the thinking that was leading them to this conclusion was governed by the ERP paradigme.

sentence, he can only think because he is. The "being" precedes the "thinking." That creates a self-reference loop. Certainty is a natural world feel, it cannot come from a third-party self-referral observation. Thinking is because of being who is because of thinking…

These examples show how the verbal space can, by limiting itself to third-party descriptions, produce logical but strange or misleading conclusions. Brain-centrism uses the extended explanatory landscape (4.12). It explains by using a mental space characteristic that these conclusions have nothing to do with the universe "out there." They are blindspots of our verbal world. Let's take one step after the other.

Self-reference and infinite regress are two faces of the same coin. To produce an infinite regress, it is necessary and sufficient to stipulate that the regressive function, when applied to an element in the series, adds a new distinct object to it. If the new object is not distinct, we have a self-referral loop.

Many self-reference sentences are innocuous; however, essential other ones need a blocking element to make them acceptable under ERP. These blocking features are unexplainable within the system describing the other elements of the series. That is because their nature differs from the other elements of the recursive series: they have no cause to be found in the same system. ERP has to change its description of reality out there to account for these unexplainable blocking elements. It does that by introducing new fill-in concepts referring to non-observable entities. We have seen a few examples like eternal gods, atoms, nothingness, or universes. Another important example of a blocking element is the notion of *"pure randomness"* that plays such a fundamental role in quantum physics[69]. Remember how

---

[69] Bell's theorem in quantum physics imply that the probabilities that arise in the descriptions. cannot be due to our ignorance of pre-existing local variables. They constitute our best possible knowledge of quantum phenomenon. However Bell's theorem can be interpreted as the non-computability of the functions describing transformations of the phase space of the quantum system. That implies that quantum randomness in the mental space is equivalent to Gödel's indeterminacy, and true randomness is another mental space blind spot. It says nothing about "out there."

Einstein opposed his whole life the idea of using probabilities in physical descriptions. Randomness is a lack of order (2.6) and orders need mental space to exist. There can be no randomness out there, Einstein's deep intuition was correct.

In the mental space, any causal series of descriptive representations is ultimately based on an unjustified bedrock of descriptive foundations. In mathematics, for instance, one has to choose axioms. These axioms are chosen based on the mathematician's intuitions and experience. However, intuition and experience do not belong to the verbal world, so in the verbal world, they are unjustified.

An infinite number of proposals are undecidable, because of Gödel's indeterminacy. For the mental space, this lack of justified foundations is not problematic, because any understanding is ultimately anchored on acquaintance knowledge and physical emotions, that means in the natural world. Gödelian information thus precedes any descriptive knowledge (3.5). This allows the mental space to make choices for no reason related to the system. In the mental space, having no cause within the explanatory system is a common thing. Most usually choices are based on emotions, feelings, and beliefs not directly related to the system that has brought up the choice. But for ERP physicalism, who limits itself to descriptive knowledge, this is problematic.

ERP physics only considers external causes as explanations (the right side of the figure (4.12.1)). It must assume that the blocking elements are at the physical level "out there." It has, for instance, to accept that there are events with no causes, *pure random* events in the universe. Brain-centrism extends the explanatory landscape to the mental space. Randomness can only be a mental abstraction, a fill-in of the verbal world. In the second volume of this book, we will examine the consequences of this remark for the interpretations of quantum physics.

We must now make an important distinction among two types of "self-referring" propositions:

Let consider the proposition: I am a sentence that includes fewer than 11 words. Once we count the words in that proposition, it will convince us of its validity. The truth of such sentences can be verified algorithmically. Though the proposition may look self-referral, it is not. The sentence does not speak about itself, but about the text that composes it, it says something about its syntax. The truth value of this proposition could change based on the manner or the language in which we phrase it. The sentence: I think that I am a proposition that includes fewer than eleven words, has the same meaning but has a unique truth value and is incorrect.

One must distinguish a sentence from the text that composes the sentence as being on two different abstraction levels. The sentence's syntax does not contain its truth value. This sounds very like what we have observed with reduction in (3.7) and has the same origin in the mental space.

In contrast, a sentence such as I am an unprovable proposition that Gödel used in his proof (4.5), is self-referral, since it refers to its semantics. The truth value of such propositions will not change if one changes the phrasing or the language in which we write them. An extensional property of a sentence is a property that does not refer to the syntax or the structure of the proposition. The truth value of a proposition and its provability are extensional properties. Propositions that address their own truth value or provability are therefore to be considered self-referential. By referring to its semantics, such sentences drive the reader out of the syntactic system. The root of undecidability and uncomputability in formal systems lies in this idea.

The human mental space can make the difference between these two types of self-referral propositions. That difference would not appear in a third party purely syntactic system. Both are abstract sentences, but the first is an abstraction about a lower level and its truth value is computable, on the contrary, the second is an abstraction about itself, and its truth value is not algorithmically deductible from the sentence, or is undetermined.

What allows the mental space to differentiate the two sentences is our sensitivity to extensional properties like understanding and feelings. This sensitivity is because of our natural world. The computationalism paradigm (5.13) that would consider feelings and understanding as algorithmic would not allow this distinction.

As we already know, we cannot find truth values in syntax, Leibniz's dream failed.

ERP by only considering Shannon's information will miss the reference to feelings and emotions as an initial primitive cause, and will not describe human behavior.

When a third-party description comes to an effectively self-referral description, we can find no algorithmic answer to the extensional properties of the sentence. Brain-centrism calls this situation a blind spot.

It is the case when one limits the explanatory landscape to "out there" (the right side of (4.12.1)). One gives up the possibility of using mental space properties or mechanisms to explain observations. One has to imagine explanations as properties of "out there."

Primitive concepts are necessarily self-referral and thus undefinable intensionally. The verbal representations of such concepts will allow building effective self-referral sentences, such as I am a liar, whose truth value will be undecidable and create mental blind spots.

This mental phenomenon explains many limits to scientific knowledge and mathematics. Gödel's incompleteness and Turing's uncomputability are blind spots. We know mental spaces escape blind spots because we do not constrain them to remain within a formal system or within the verbal world. They can take a decision triggered by the natural world. A decision will allow stopping the self-referential nature of a regress by introducing an additional element.

At the underlying level of the physical space, the hybrid nature of the brain information system that Miguel and I have described in the Relativistic Brain theory represents this mental capacity. Turing had already figured out this mechanism in his paper of 1939 by describing the Oracle machine. (4.10)

"Out there," there are no self-referral propositions, no infinite regress, and no blind spots. Asserting that they happen because of the "nature" of "out there" cannot be satisfactory. Nature is not blind; our representations are.

We base our ancestral individuations of natural entities on information collected through our sensory system. But at the quantum level, where only the extended definition of existence can be used, our individuations are mathematical.

* *

6.11 Intelligence, Artificial Intelligence

The word intelligence is a high-level abstract concept only definable intensionally. It describes a feeling of the natural world. We know that reason and logic are not the sources of intelligence, they are rather one of its products. (4.6 to 4.10.) Originally, intelligence is a feeling we have about someone else. The feeling we can rely on his judgment or his understanding. Or even perhaps in his fairness or his creativity. The feeling of intelligence developed when human societies organized. Just like the idea of time, we did not separate the idea of intelligence from many other mental characteristics. When we gave a name to intelligence, we "objectified" it and wanted to measure it. As always, the danger is that we call intelligence what we are measuring and create an antinomy with the "feeling of intelligence." Intelligence is an umbrella term that covers a variety of mental abilities. These abilities are all interconnected, interacting, recombining in different proportions for each of us, in a manner that varies with the mental state and the environment. One can recognize various fragmented aspects such as survival, understanding, feeling of time, memory, creativity, intuition, feelings, mental patterns, imagination, problem-solving, mental attitudes, survival capacities. Although we decompose our feeling of intelligence, we know that these aspects, and many others, cannot be separated one from another. Fragmenting and isolating some aspects without the others can lead to confusion and mistakes.

Without survival, for instance, most aspects of intelligence disappear. If aging and death aren't to be expected, the very meaning of most aspects would be different. Without the feeling of time, our entire approach to solving problems would be different.

By giving a name to this complex and variable combination of feelings, capacities, characteristics, and tendencies of the mental space, we individuate "something," treating it like "objective" reality. We also neglect aspects that may be sometimes critical.

This description of intelligence involving many Gödelian information aspects excludes the possibility for intelligence to be algorithmic and makes the term Artificial Intelligence, AI, sound strange. How could "something" woven in the fabric of our organic brain be reproduced artificially? How could pure syntax generate it?

John McCarthy invented the term Artificial Intelligence, then a young Assistant Professor of Mathematics, to promote his summer workshop in 1956 at Dartmouth College in Hanover, New Hampshire. A few years before McCarthy had published with Claude Shannon a book entitled "*Automata Studies*," who had little impact. The genius of McCarthy was to understand the importance of a new name; Automata studies would become *Artificial intelligence* for the occasion. This was a great "marketing idea" and we now considered the Dartmouth workshop as the launchpad of Artificial Intelligence.

Huge personalities such as John McCarthy, Marvin Minsky, Julian Bigelow, John Nash, Nathaniel Rochester, Ray Solomonoff, and Claude Shannon were among the promoters and participants at the 1956 Dartmouth workshop. All these scientists became famous and recognized leaders in AI research for decades.

Over 13,000 experts took part at the world's leading academic AI conference in 2019 NeurIPS in Vancouver. Blaise Aguera y Arcas an authority in computer vision declared:

"Deep learning has rapidly knocked down some long-standing challenges in AI, but it doesn't immediately seem well suited to many that remain... The models that we have learned how to train are about passing a test or winning a game with a score, [but] so many things that intelligence does, aren't covered by that rubric at all."

At that same conference Yoshua Bengio, that is credited to be a founder of deep learning said:

"We have machines that learn in a very narrow way. They need much more data to learn a task than human examples of intelligence, and they still make stupid mistakes."

Today's methods combine both connectionism and symbolic approach. They try to mimic more closely the processes of the human brain that apply both ways and other alternatives.

Melany Mitchell, professor of computer science and author of *"Complexity, a guided tour,"* says:

"During the first several decades of AI research, many scientists believed they would be able to manually program computers with this model-based knowledge of the world, and the programs would be able to use logic to understand the situations they encountered, make plans, and deduce the likely future. However, this knowledge engineering project was doomed. Our mental models are too vast, too filled with unconscious knowledge, and too enormously interconnected to be captured manually and with rigid logic. So-called knowledge-based approaches to AI turned out to be brittle—unable to cope outside narrow domains —and were largely abandoned."

We have overcome many of the nursing problems, and our technologies are here to stay. Computer scientists also have access to much more powerful machines, allowing deeper algorithms to run. We discover that more tasks than we believed can be mechanized, and AI applications are growing in unsuspected fields.

But although the advances in AI have been gigantic, we should not forget that it has little to do with human intelligence. It remains an automation tool, branded "intelligence" for marketing reasons. Now many researchers are putting a lot of effort into imitating human intelligence, branded AGI Artificial General Intelligence.

This is what the physicist Sabine Hossenfelder says on her blog:

Physicists avoid the term "Artificial Intelligence" not only because it reeks of hype, but because the analogy to natural intelligence is superficial at best, misleading at worst. True, the current models are loosely based on the human brain's architecture. ... The current algorithms heavily rely on humans to provide suitable input data. They do not formulate own goals. They do not propose models. They are, as far as physicists are concerned, but elaborate ways of fitting and extrapolating data.

Using the same name, intelligence, for human activity and a computer makes limited sense and can be very confusing and antinomic. By individuating "intelligence" in a limited manner, and neglecting many fundamental characteristics of humans, brain-centrism considers that ERP is embarking us in a dangerous direction. The danger comes from the "imitation fallacy" we described in (5.14). The mental space can confuse original and imitation and establish the imitation as the new norm, destroying profound and essential human characteristics.

* * *

VII: WE HUMANS

> *Because the stage of our reality is the mental space, separating science and humanism is a lure, just as isolating knowing from being. All aspects of inner and outer knowledge contribute to who we are and how we will develop. Just as true morality comprises being moral and not only knowing about righteousness, true generosity comprises being generous, and genuine kindness comprises being kind, not of cultivating these qualities, hoping to gain something. The imitation fallacy with the use of technology is taking incredible proportions, it has destroyed credibility for marketing and show off. Greed has been institutionalized. A new enlightenment age is necessary if we want to survive.*

7.1 The imitation fallacy

Alan Turing had proposed in his 1950 paper, "*Computing Machinery and Intelligence*" a game that he named the imitation game. In this game, a human evaluator would judge the capacity of a computer to imitate humans. Two parties hold conversations in natural language, a human and a correspondent. That correspondent could be a machine, designed to generate humanlike responses. The human evaluator has to guess if he is speaking with a machine or another human. This game was rebaptized Turing test and serves as a benchmark for artificial intelligence. If the evaluator cannot reliably tell the machine from the human, the machine is said to have passed the test.

Turing had previously expressed the idea that the intelligence of a machine was limited. It could imitate what a human does when he is calculating. That means when he is applying an algorithm. (4.7). In his 1950 paper, he argues against all objections to the proposition that "machines can think." This paper would have an amazing influence although Turing did not explicitly state that his game could be used as a measure of intelligence.

John McCarthy introduced the term Artificial Intelligence to promote his summer workshop in 1956 at Dartmouth College. (6.11) It is an unfortunate designation, as intelligence is not definable by description (6.10), and it does not apply to a machine.

Today computers play chess with better results than humans; however, computers don't play the game the same way humans do. They use different routes to achieve their result. (5.8.1) If one individuates the game of chess as the rules of chess, then humans and computers play the same game. However, if one individuates the game of chess as the mental activity associated with the rules, then they play different games. ERP, however, will not recognize the individuation of the game extending to the mental spaces of the players. In certain cases, as we will see below, extended individuation may be crucial.

An imitation can only be identical to the original it imitates if one chose a restricted individuation.

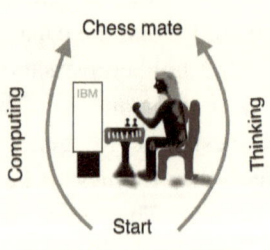

Figure 7.1.1 : Are they playing the same game?

ERP by considering that causes can only belong out there, on the right side of the figure (4.12.1) will neglect causes that manifest in the mental space. In our chess example, ERP would concentrate only on the "objective" rules of chess. (7.1.1) Brain-centrism considers that the game has been designed by humans and for humans, it consists not only in the algorithmic rules but also include the whole process humans use to play it, starting in the mental space, on the left on the figure (4.12.1). The entire thinking process is then part of the game, not only the result. The pleasure in playing chess is in the thinking process, not in the rules.

This is the case for many human activities, the path to get there is more important than the arrival line. Nobody wants to speed a piano concerto or enjoying an excellent wine, or a dance with your partner, just to "get it done" as fast as possible.

We have designed those activities for humans; we enjoy the complete process. When feelings are involved, the goal of an activity is not to finish it as fast as possible. "Objective measuring" makes little sense. The ERP materialist vision focuses on what is measurable, however, the important things in our lives mental. Most are Gödelian and thus not measurable.

We cannot submit composing, painting, thinking, understanding, playing, and any activity requiring creativity to the constraints required by economic productivity. Only algorithmic activities can be rationalized and subject to mass production.

By remaining in the digital world, one cannot distinguish an original from a copy. It is only possible in an analog world where individuation extends beyond what can be measured. (4.12) In a digital world were one abstracts only Shannon information any copy is an original. The Shannon information of two objects can be equivalent but for our human mental space trigger very different Net(R)'s and thus different feelings and reactions.

Christies sold a copy of the first edition of Isaac Newton's Principia dated 1687 for $1.5m. It contains the same Shannon information as the free ebook you can get online. To understand the difference, one has to consider the left side of the figure (4.12.1). By considering only the Shannon information content of the book, the measurable part, we can see no difference between an original edition and a copy paste one.

Let's imagine a perfect imitation of a Van Gogh painting, indistinguishable from the original using the most sophisticated instruments. Every measurable criterion would tell us it is a Van Gogh painting. However, we have again restricted our individuation to what is measurable. You may be amazed by the imitation, but knowing it's not Van Gogh who painted it, you would not pay the same price. For us humans, the important part lies in the natural world. And this part is not computable.

An imitation, even perfect is something different than the ground-level original. It's about the original. We may not have the information, believe it's original, and treat it as such. We are then at risk of one day waking up to the truth.

This is the essence of what we call the *"Imitation fallacy."* The imitation fallacy restricts the individuation of a phenomenon in such a way that one confuses an original with an imitation.

ERP can cause imitation fallacies because it restricts its explanatory landscape to the right side of the figure (4.12.1). Brain-centrism proposes, when necessary, to extend the individuation range and avoid confusing simulated maps with the original territory.

"People can believe in the multiverse all they want—but it's not science." Says Sabine Hossenfelder, physicist at the Frankfurt Institute for advanced studies and renowned blogger and author. She adds:

"Theoretical physicists used to explain what was observed. Now they try to explain why they can't explain what was not observed. And they're not even good at that."

Here is a reader's comment I found recently on a physics website, that illustrates how physics is at risk of deriving from science to a belief system:

The last pronouncement, about visual evidence of the black hole phenomenon, occurred on Wednesday. A black hole is now a scientific truth because of three main reasons: One, people with great authority, who have monopoly over a narrow field of study, have said so after an arcane process that is widely believed to be very rigorous; two, other people like them have endorsed it; three, most people in the world, including scientists in other fields, do not have enough information to challenge the assertion. Also, the journalists, who usually seed doubt in the minds even in areas like genetics and climate do not challenge scientists on theoretical physics.

Erwin Schrödinger, who was a very independent personality, wrote very critically about what he called "the Barbarian scientist":

"He reaches the point where he proclaims it a virtue not to take any notice of all that remains outside the narrow domain he himself cultivates, and denounces as dilettantism the curiosity that aims at the synthesis of all knowledge."

The intrinsic value of science is trust. The public must trust what scientists are presenting as our best effort to approach truth. Imitating a friendship behavior, even perfectly, does not mean I am your friend. Imitating love is not love. Imitating sincerity is not sincerity. Imitating a feeling cannot be that feeling, because it implies an intention that feelings cannot have. Imitation fallacy can dissimulate everywhere. People use it to "make-believe" and find some advantage in presenting themselves as they are not.

The judiciary systems recognize that not only the "objective" facts are crucial but also the intentions of the indicted person. Justice recognizes that our intentions and the state of our mental space are also important. One cannot neglect what happened mentally to evaluate a situation where humans are involved. Neglecting it is suppressing what makes us human, it is mechanizing.

For brain-centrism, they are no situation not involving humans. ERP creates this illusion because it uses third-party descriptions, brain-centrism considers that any description is a human description.

Why is the bust of Nefertiti, coasted 1312 BC and exposed in Berlin, different for a perfect imitation of it? Because our mental space is involved, not only the physical matter of the statue. We attach to the "original" a history and complex feelings relating us to our past. An imitation can be perfect. It does not generate quite the same mental attachments. Millions, since its discovery in 1912 have traveled to visit the original Nefertiti. No one travels to visit a copy, regardless of its perfection. Our human life happens in our mental space, and the characteristics of this space dictate our decisions and actions.

We admire the sincerity of younger children, their nativity and lack of intentions, of hidden agendas, or calculated reactions. They act with no imitation fallacy. The "mental route" leading to a behavior is important because it says a lot about the feelings, the intentions, the purpose, the history, and the feelings behind an action.

To survive we have learned how to calculate, hide and be, "clever." We are trained to use our verbal world and to hide our natural one. Imitation has conquered the world and occupied the mental spaces. We spend more time pretending or "imitating" whom we are not, rather than time discovering who we are. We know that so well that we have learned "not to trust" to protect ourselves. The social and economic cost is enormous.

Hidden agendas, vested interests, and corruption are results of the imitation fallacy conjugated with an economy that rewards the result and neglects the route used to get it.

For many practical purposes, imitations are necessary and convenient. Imitations become fallacies when they are presented as equivalent to originals, or when one hides that something is a copy.

The ERP limited individuations have driven us to dangerous visions like behaviorism that focuses on "objectively" observable behaviors discounting mental intentions. With new technologies, limited individuations could lead us to further destructive absurdities. The very idea of human-level Artificial General Intelligence (AGI) is an imitation fallacy. Because it pretends not only to imitate the verbal world but also the natural world and its Gödelian feelings.

Many thinkers and entrepreneurs have reported that AGI can be dangerous. They are correct but often present the wrong reasons. AGI is a theoretical vision, impossible to realize on digital machines. The danger is that they might present one day to us something as an AGI. It will only be some stupid sophisticated imitation machine, based on narrow individuation of human intelligence and feelings. We will be instructed to believe it as hyper-intelligent, although it has no human emotions. We will forget it is a tool, a machine, its narratives might fool us. We will do our best to adapt to them. Our dogs, whose representations are based more on smell and much less on narratives, will not.

During the last 20 years, they have educated us to deal with machines. We have learned to overcome our frustration when confronted with their stupidity. We spend hours trying to get an answer that a human could give us immediately. They instructed us to call that progress. The public is doing now for free what the companies were doing before. They have mechanized us to save money, and we have accepted. With AGI, progress will speed up.

* *

7.2 Fighting emptiness by learning

Mammals can get bored, especially in captivity, when their environment does not show sufficient complexity to keep their mental space stimulated. In the wild, it does not happen. It never happened to our savanna man. But it happens to us and our children today. Boredom can develop when we overcome a threshold of security, and no new challenge keeps us busy. It is an unpleasant feeling, showing that our mental space, driven by curiosity, is hungry for new challenges and new stimulations. With no natural stimulations, it will concentrate on artificial entertainment.

Dr. Sandi Mann, a psychology lecturer at the University of Central Lancashire, explains:

"The more entertained we are, the more entertainment we need to feel satisfied. The more we fill our world with fast-moving, high-intensity, ever-changing stimulation, the more we get used to that and the less tolerant we become of lower levels."

The Danish philosopher Søren Kierkegaard (1813–1855) described boredom as a feeling of emptiness. It is not, according to him, the absence of stimulation that triggers boredom, but an absence of meaning.

People get bored, not only because of lack of stimulation but also because their mental space does not find a deeper meaning to what they experience. Boredom would be a feeling inviting us to reorganize our mental space to make sense of the world we are experiencing, or change our environment. We need stimulations, but we also need to gain meaning through what stimulates us. We need to generate new Gödelian information.

Entertainment can overcome boredom, but it would not solve the problem. It would only increase our future need for more entertainment. Only a change in our attitude towards the world and our self-image could protect us from boredom. Switching on the TV or a video game is so much easier than questioning ourselves and broadening our individuations. Never getting bored is thus a matter of education on the "being line."

The way certain humans consider their life does not give them a second to feel bored.

"There are only two ways to live your life. One is as though nothing is a miracle. The other is as though everything is a miracle." Albert Einstein.

Do we see ourselves as nothing or as miracles? The progressive shift towards third-party perspectives, the immensity of the universe, the growing mechanization of our human society, and its apparent absurdity, leave us wondering where we stand. The world could discourage us when we open our eyes. Discovering who we are is no more sufficient, we must make choices and decide who we want to be. We must decide what will be our most important dreams and strive for them. Conforming and adapting is not sufficient. We may feel so weak and insignificant, that such decisions may be the most crucial but fundamental factor determining our personal and collective future.

This future starts during the 100 milliseconds we dispose to interfere and change our immediate usual response and start thinking. (3.12). Thinking is the only guide we have to confront the complexities we are facing, and nature has only given us a brief time to start the process. One hundred milliseconds, often blurred by a permanent flow of distracting events designed to capture our attention. Choices and opportunities come to our minds when we think. They don't come as ready-made thoughts, but as developing feelings that confront us with situations and people who can share these feelings. When we open up our mental space, decisions come, and "out there" also opens up.

Today, they embarked us in a generalized, all-encompassing model of lingering stimulations and rewards that, although they are entertaining, erode our thinking opportunities and limit our attention span. The average western citizen receives more information in a week than he would have received in a lifetime a few hundred years ago. That leaves him powerless to absorb, digest, and decide. It is then easier for him to remain a spectator. This attitude favors the undoubting narcissistic individuals in search of power.

Salespeople, politicians, advertisers, people with vested interests, who tell us what to think about, how to think about it, and want us to follow their suggestions rather than develop our thoughts invade us. Since young, they train us to conform and to follow.

The continuous flow of information concerning subjects we cannot control is a prison and has a terrible effect on our stress levels. It conditions us to believe that there is no other solution to live our lives than to follow and conform. Brain-centrism teaches us we have control over our thought, we can at every moment choose to give up this control to others or not to do so. There are aspects of our lives where we feel we can act, and those are the aspects we should concentrate on. Giving too much attention to the flow of information, we cannot act on, will only bring stress and make us feel useless.

Like the Sophists 25 centuries ago, some people and organizations today have understood how our mental space works. They use this knowledge to become our exclusive thinking guides, and invade our mental space, appealing to its weaknesses, rather than its strengths. We have built a society that uses the information to teach us how not to think, how not to realize who we are or where we are.

For brain-centrism, some distance from this continuous flow is necessary from time to time to realize what we can do if we catch these 100 ms. It is not so easy for many of us to find time and establish this distance. Our mental patterns will force us back to our usual behaviors. But only thinking by ourselves, imagining, creating our beliefs can allow us to find meaning and decide who we want to become. No ERP third-party perspective can reveal what we can only ourselves discover. Our truth is not in the verbal world, it's in our natural one.

* *

7.3 Conclusions

> *In a time of increasing darkness, we must respond increasing the light.*
> Kabbalah

Over the years, many brilliant thinkers have become my heroes. For some of them, they even became my friends. When I felt lonely, I could come back to them, and they would teach me something. Amazing things that nourished my soul, stimulated my imagination and made me feel proud to belong to the human species. A pride one does not grasp in day-to-day life. Gödel, Poincaré, Russell, Aristotle, Democritus, Mach, Leibniz, are heroes on top of my list. But these last ten years Albert Einstein made it to the very top, even before our common friend Kurt Gödel.

I am certain I don't understand all his thinking. Reading his articles and his books, imagining his conversations with Tagore, with Niels Bohr, and with Kurt Gödel, I could appreciate how he matured as a man and as a scientist. I could feel his doubts and his persistence in pursuing an idea. I could appreciate his sense of vanity and his humor when he joked, declaring he had become a relic deposited in a museum. This guy was my absolute best friend. I would have loved discussing brain-centrism with him.

Here is another citation that shows the deep humanity that connects him, millennia apart to our cave painter Shaman:

"To sense that behind anything that can be experienced there is a something that our mind cannot grasp and whose beauty and sublimity reach us only indirectly and as a feeble reflection, this is religiousness. In this sense I am religious."

Einstein was a subtle philosopher. He chose realism as one study German instead of French. ERP was his dogma. Within the framework of the ERP dogma, he could bring his contributions.

He knew well that this was a choice, only a working hypothesis to develop his theories without having to care what was happening in the mind of an observer.

In July 1930, in Berlin, Einstein met several times with the Indian poet and philosopher, the Nobel prize Rabindranath Tagore.

This is an excerpt of the first encounter between two men of different cultures and backgrounds:

Einstein: I cannot prove scientifically that Truth must be conceived as a Truth that is valid independent of humanity; I believe it firmly. I believe, for instance, that the Pythagorean theorem in geometry states something that is approximately true, independent of the existence of man.

The usage of "approximately true" shows that Einstein is speaking as an ERP physicist. But the referral to a mathematical truth indicates he is also Platonist, a necessity we mentioned earlier for any materialist. The discussion continued:

Tagore: Yes, one eternal entity. We have to realize it through our emotions and activities. We realized the Supreme Man who has no individual limitations through our limitations. Science is concerned with that which is not confined to individuals; it is the impersonal human world of Truths. Religion realizes these truths and links them up with our deeper needs; our individual consciousness of Truth gains universal significance. Religion applies values to truth, and we know this truth as well through our own harmony with it.

Tagore introduces the idea of realizing the truth through our human limitations. He distinguishes impersonal human truth studied by science from our consciousness of truth. In both cases, humans are at the center.

Einstein: Truth, then, or Beauty is not independent of Man?

Tagore: No.

Einstein: If there were no human beings anymore, the Apollo of Belvedere would no longer be beautiful.

Tagore: No.

Einstein: I agree with regard to this conception of Beauty, but not with regard to Truth.

Tagore: Why not? Truth is realized through man.

Einstein: I cannot prove that my conception is right, but that is my religion.

People often asked Einstein: "how do we know that objects in our models correspond to real things in the actual world?" He used to say with a big smile that there are questions that we cannot answer.

Esther Salaman was a Russian Jewish physicist, she was a student of Einstein in Berlin and became his friend. She wrote her final memoir on Albert Einstein. During a walk together in 1925, Einstein declared:

"I want to know how God created this world. I'm not interested in this or that phenomenon, in the spectrum of this or that element. I want to know His thoughts; the rest is just details."

This was his deep idea, the one that would occupy him until he died in 1955: The Theory of Everything, the ultimate "unification," the ultimate abstraction.

His thoughts, as he expressed them to Esther, clarify that his mind went far beyond ERP. He deep hoped to penetrate the Platonic realm that accompanies physicalism. When asked about his religious beliefs, he came up with different answers: agnostic, pantheistic, religious non-believer. But he often specified that he did not believe in a God who interferes in human affairs. His convictions were closer to those of Baruch Spinoza: "a god who reveals himself in the harmony of nature."

At the IAS in Princeton, he used to walk back home with his friend Kurt Gödel. One could often see these two very different men wander a few steps and stop to discuss or think. In his latter days, Einstein declared that he came to the institute to have the privilege *"to be able to walk home with Gödel."*

Gödel was a convinced Christian and believed in a personal god, that interfered in human affairs. In his book *"Tomorrow's God: The Hebrew Lord in an Age of Science,"* Robert N. Goldman narrates a discussion he had with Gödel after Einstein's death in 1955. In that discussion, Gödel confirmed Einstein believed that science could never explain consciousness:

"The methods of science lead away from the life world. Einstein believed that space and time are human constructs, but reality was something deeper."

I have been very fortunate to have these scientific friends. I recognize now how "religiously" grateful I feel to have the chance and the curiosity to study their legacy. How privileged I am to be sensitive to their sincerity and the unlimited devotion they had to their work. I admire their successes. But I also feel for their failures, their doubts, their revolt, and their persistence. Getting older, I suppose we all feel growing disgust for power plays, imitations, arrogance, vested interests, and marketing strategies. My heroes were sincere; they were not playing around to market their personalities. They were hopeful because they recognized the fragility of our human condition, their fight came from recognizing our weakness. From their fight emanates the very fabric of what keeps us alive: hope.

Some of their ideas are like pieces of art; we can start by loving them for the inspiring music they play in our hearts, and the order they put in our brain. But, one day, we realize that these pieces of art were not always there. A human being created them, a person just like us, who doubted, just like us, and transformed his doubts into scientific ideas just as monumental as the great Gizeh Pyramids. Be it pyramids or ideas, they connect us to humans that lived, strived, hoped, loved, and dreamed things so important for them they would never stop. Only these connections through the centuries can keep us alive by broadening our foundations. Looking backward, one can see the fulfillment, not the struggles and the pains. Those things evaporate, hope remains.

How many billion people have lived on earth for only a few dreamers to paint an immortal symphony? How many billion stars in the universe were necessary for intelligent life to emerge?

When you look at a painting, let's say a Picasso, you can only appreciate it, it can only fill your soul if you are ready to receive it if you have the culture. That is because you represent this painting in association with memories and emotions, what we called Gödelian-information. What you will feel at that moment is yours, it's in your brain. But it connects you to other humans who can experience similar music. This connection is not about explaining something, it's deeper and only reserved for humans.

You can try to put words on it and explain the painting, but your explanations will only be a bad linear projection. You experience a multidimensional space; it possesses multiple attributes simultaneously, evokes important feelings and memories. All in an in-dissociable bundle, in unity, that you lose when you use a linear language. Explain a Mozart symphony, all you get is words, the soul got lost.

But the one who expressed it the best is Picasso himself. He took part at an exhibition of his paintings in Cannes in the 1960s, and a journalist asked him:

Mr. Picasso, can you explain to me what this painting means?

Picasso stayed silent for a long moment, contemplating his abstract painting, like dreaming, then he replied:

Sir, if I could explain it, I would not have painted it.

Essential words are the ones we cannot define. Essential events cannot be programmed. What matters the most gets diluted when it is expected. Profound discoveries are never part of a research program.

Genuine hope does not come from promises, we do not find it through greed, nor reasoning. The profound nature of things will always escape the verbal world, but seeking it makes us human. The most challenging mathematical questions are the unsolved ones. Most remarkable encounters happen by chance. The most frightening sheet of paper is the empty one, but once we overcome fear, it can become our best confidant.

We are at our best as human beings when we can act and feel spontaneous when we are not following rules that pre-program us.

We should not surrender to fear; our mental space can try to fight it. We should not always go for the safest solution; we should continue exploring and not swallow predigested explanations.

What a tremendous difference if you do something because you want to do it and if you do it because you are obliged. We humans can do things because we want to, for no other reason. They, robots, follow their programming.

By trying to identify us with computers, and treating us as if we were machines, I am afraid that our society is killing the best part of what we are: Independent but social individuals, each with their unique personality, able of imagination, of creation, able to show a huge empathy for our brother humans, able of hope, marvel, and dreams.

We should not let ourselves be managed like machines. We humans understand, they don't. We need to understand to survive. We need to allow our primitive sources of emotions to continue thriving. To be alive in our minds and regulate our thoughts. We are not digital machines.

As we discuss many times in this book, we have reached a crossroads, a bifurcation point where new paradigms may emerge. A new image of our planet, of humankind, of our position in the universe, is now ready to grow.

The problems that we have created will not be solved by accelerating the same old methods. We are reaching the end of a Kuhn "normal" phase, in scientific research, in technology, but also healthcare, and sociology. But perhaps above all, we will have to imagine the values that can guide us towards a new enlightenment era.

Copernicus put the Sun at the center of the planetary system, asserting that there is no privileged place to observe the universe. This was a lesson of humility. Now brain-centrism puts us back in the center of the universe generated by our mental space but loses the universe out there. This should provide even more humility.

Our history of intervening in complex systems is not glorious. Humans are all too willing to interfere with orders that we barely understand. We want to improve them, but according to our standards. When we embark on such projects, our intentions might be good, but the effects can still be terrible down the road. We sometimes proceed with a worrying arrogance.

Human history looks like a long succession of disasters, one after the other, wars, plagues, ideologies, slavery, violence. When you sum all this up, we have, in a few thousand years, made incredible progress. A few hundred years ago, most of humanity lived in slavery, 1% possessed 99% of all the richnesses, they reserved higher education to an elite, and authoritarian systems dominated the planet.

From the brain-centric perspective, ERP is an illusion, a "make-believe" so powerful that it is quasi-unescapable once installed in a mental space. Illusions have populated the mental space with a world of "material objects" that have no "material existence." Man-made obstacles that have become "realities." Objects that one cannot even see, or touch, or smell, but that dictate our behaviors. The belief in illusory objects is so strong that the infected mental spaces can destroy themselves and others for the sake of these illusions. Because of self-reinforcement and self-confirmation, because of the very nature of the mental space, these materialized illusions appear as low-level abstractions while they only are materializations. Economy, religions, social constructs, ideologies, but also science are full of these illusory and destructive objects.

By not distinguishing low-level representations from higher-level materialized abstractions, we have built a reality that puts us at risk of self-destruction. We individuate mental entities that have no existence out there as existent and give them the power to rule our lives, but also sometimes to destroy them.

Therefore, brain-centrism asserts that we need a new enlightenment age, that would more deeply review our thinking processes. The 18th-century enlightenment made a big step by giving priority to reason and experimentation and recognizing that humans must be central to any societal organization. We now need a further step that would recognize the illusory nature of ERP in science, in the economy, and the organization of human societies. It does not claim human rights as ERP facts; the new enlightenment should dig deeper into the characteristics of the human mental space and its unlimited varieties. It should recognize who and where we are, we humans.

Rousseau believed that the human soul was good and Voltaire that it was bad. I trust we have the choice at every second. Those 100 milliseconds given to us to interfere and change any immediate mechanical reaction, Libet's 100 milliseconds, is the place where our freedom hides. Our opening to the world of thinking and discovering. That is the point where everything can change. Nature has given us the opening but also the tool, our mental space, the true creator of our universe.

* * *